PROGRAMMING
KOTLIN® APPLICATIONS

W0246342

Programming Kotlin® Applications

Programming Kotlin® Applications

BUILDING MOBILE AND SERVER-SIDE APPLICATIONS WITH KOTLIN

Brett McLaughlin

A Wiley Brand

for Leigh, as always, my person

ABOUT THE AUTHOR

BRETT MCLAUGHLIN has been working and writing in the technology space for over 20 years. Today, Brett's focus is squarely on cloud and enterprise computing. He has quickly become a trusted name in helping companies execute a migration to the cloud—and in particular Amazon Web Services—by translating confusing cloud concepts into a clear executive-level vision. He spends his days working with key decision makers who need to understand the cloud as well as leading and building teams of developers and operators who must interact with the ever-changing cloud computing space. He has most recently led large-scale cloud migrations for NASA's Earth Science program and the RockCreek Group's financial platform. Brett is currently the Chief Technology Officer at Volusion, an ecommerce platform provider.

ABOUT THE TECHNICAL EDITOR

JASON LEE is a software developer happily living in the middle of the heartland. He has over 23 years of experience in a variety of languages, writing software running on mobile devices all the way up to big iron. For the past 15+ years, he has worked in the Java/Jakarta EE space, working on application servers, frameworks, and user-facing applications. These days, he spends his time working as a backend engineer, primarily using Kotlin, building systems with frameworks like Quarkus and Spring Boot. He is the author of *Java 9 Programming Blueprints*, a former Java User Group president, an occasional conference speaker, and a blogger. In his spare time, he enjoys spending time with his wife and two sons, reading, playing the bass guitar, and running. He can be found on Twitter at `twitter.com/jasondlee`, and on his blog at `jasondl.ee`.

ACKNOWLEDGMENTS

I USED TO WATCH MOVIES AND STARE in amazement at the hundreds of names that scrolled by at the end. How could so many people be involved in a single movie?

Then I wrote a book. Now I understand.

Carole Jelen is my agent at Waterside, and she replied to an email and picked up the phone at a time when I really needed someone to help me find my way back into publishing. I'm incredibly grateful.

On the Wiley side, Brad Jones was more patient than he ever should have been. Thanks, Brad! Barath Kumar Rajasekaran handled a million tiny details, and Pete Gaughan and Devon Lewis kept the train on the tracks. Christine O'Connor handled production, and Jason Lee caught the technical mistakes in the text that you wouldn't want to stumble over. Seriously, Jason in particular made this a much better book with his keen eye.

As usual, it's an author's family that pays the highest price. Long days, more than a few weekends and evenings, and a constant support keep us going. My wife, Leigh, is the best, and my kids, Dean, Robbie, and Addie, always make finishing one of these a joy.

Let's do brunch, everyone! Mimosas and breakfast tacos are on me.

—BRETT MCLAUGHLIN

CONTENTS

INTRODUCTION

For decades, the Java programming language has been the dominant force in compiled languages. While there have been plenty of alternatives, it's Java that has remained core to so many applications, from desktop to server-side to mobile. This has become especially true for Android mobile development.

Finally, though, there is a real contender to at least live comfortably beside Java: Kotlin, a modern programming language shepherded by JetBrains (www.jetbrains.com). It is not Java, but is completely interoperable with it. Kotlin feels a lot like Java, and will be easy to learn for developers already familiar with the Java language, but offers several nice improvements.

Further, Kotlin is a full-blown programming language. It's not just for mobile applications, or a visual language that focuses on one specific application. Kotlin supports:

➤ Inheritance, interfaces, implementations, and class hierarchies

➤ Control and flow structures, both simple and complex

➤ Lambdas and scope functions

➤ Rich support for generics while still preserving strong typing

➤ Idiomatic approaches to development, giving Kotlin a "feel" all its own

You'll also learn that while Kotlin is a new language, it doesn't feel particularly new. That's largely because it builds upon Java, and doesn't try to reinvent wheels. Rather, Kotlin reflects lessons that thousands of programmers coding in Java (and other languages) employ on a daily basis. Kotlin takes many of those lessons and makes them part of the language, enforcing strong typing and a strict compiler that may take some getting used to, but often produces cleaner and safer code.

There's also an emphasis in Kotlin, and therefore in this book, on understanding inheritance. Whether you're using packages from third parties, working with the standard Kotlin libraries, or building your own programs, you need a solid understanding of how classes interrelate, how subclassing works, and how to use abstract classes along with interfaces to define behavior and ensure that behavior is implemented. By the time you're through with this book, you'll be extremely comfortable with classes, objects, and building inheritance trees.

The Kotlin website (kotlinlang.org) describes Kotlin as "a modern programming language that makes developers happier." With Kotlin and this book, you'll be happier *and* more productive in your Kotlin programming.

WHAT DOES THIS BOOK COVER?

This book takes a holistic approach to teaching you the Kotlin programming language, from a beginner to a confident, complete Kotlin developer. By the time you're finished, you'll be able to write Kotlin applications in a variety of contexts, from desktop to server-side to mobile.

WILL THIS BOOK TEACH ME TO PROGRAM MOBILE APPLICATIONS IN KOTLIN?

Yes, but you'll need more than *just* this book to build rich mobile applications in Kotlin. Kotlin is a rich language, and while there are books on all the packages needed to build mobile languages, this is fundamentally a book on learning Kotlin from the ground up. You'll get a handle on how Kotlin deals with generics, inheritance, and lambdas, all critical to mobile programming.

You can then take these concepts and extend them into mobile programming. You can easily add the specifics of Android-related packages to your Kotlin base knowledge, and use those mobile packages far more effectively than if you didn't have the fundamentals down.

If you are anxious to begin your mobile programming journey sooner, consider picking up a book focused on Kotlin mobile programming, and hop back and forth. Read and work through Chapter 1 of this book, and then do the same for the book focused on mobile programming. You'll have to context switch a bit more, but you'll be learning fundamentals alongside specific mobile techniques.

This book covers the following topics:

Chapter 1: Objects All the Way Down   This chapter takes you from getting Kotlin installed to writing your first Kotlin program. You'll learn about functions from the start, and how to interact with the command line through a not-quite "Hello, World!" application. You'll also immediately begin to see the role of objects and classes in Kotlin, and refine your understanding of what a class is, what an object is, and what an object instance is.

Chapter 2: It's Hard to Break Kotlin   This chapter delves into one of the distinguishing features of Kotlin: its rigid stance on type safety. You'll learn about Kotlin's types and begin to grasp choosing the right type for the right task. You'll also get familiar with val and var and how Kotlin allows for change.

Chapter 3: Kotlin Is Extremely Classy   Like any object-oriented language, much of your work with Kotlin will be writing classes. This chapter digs into classes in Kotlin and looks at the basic building blocks of all Kotlin objects. You'll also override some functions and get deep into some of the most fundamental of Kotlin functions: equals() and hashCode().

Chapter 4: Inheritance Matters   This chapter begins a multichapter journey into Kotlin inheritance. You'll learn about Kotlin's constructors and the relatively unique concept of secondary constructors. You'll also learn more about the Any class, understand that inheritance is truly essential for all Kotlin programming, and learn why writing good superclasses is one of the most important skills you can develop in all your programming learning.

Chapter 5: Lists and Sets and Maps, Oh My!   This chapter moves away (briefly) from classes and inheritance to add Kotlin collections to your arsenal. You'll use these collection classes over and over in your programming, so understanding how a `Set` is different from a `Map`, and how both are different from a `List`, is essential. You'll also dig further into Kotlin mutability and immutability—when data can and cannot change—as well as a variety of ways to iterate over collections of all types.

Chapter 6: The Future (in Kotlin) Is Generic   Generics are a difficult and nuanced topic in most programming languages. They require a deep understanding of how languages are built. This chapter gets into those depths, and provides you more flexibility in building classes that can be used in a variety of contexts than possible without generics. You'll also learn about covariance, contravariance, and invariance. These might not be the hot topics at the water cooler, but they'll be key to building programs that use generics correctly, and also level up your understanding of inheritance and subclassing.

Chapter 7: Flying through Control Structures   Control structures are the bread and butter of most programming languages. This chapter breaks down your options, covering `if` and `else`, `when`, `for`, `while`, and `do`. Along the way, you'll focus on controlling the flow of an application or set of applications all while getting a handle on the semantics and mechanics of these structures.

Chapter 8: Data Classes   This chapter introduces data classes, another very cool Kotlin concept. While not specific to only Kotlin, you'll find that data classes offer you a quick and flexible option for representing data more efficiently than older languages. You'll also really push data classes, going beyond a simple data object and getting into constructors, overriding properties, and both subclassing with and extending from data classes.

Chapter 9: Enums and Sealed, More Specialty Classes   This chapter introduces enums, a far superior approach to `String` constants. You'll learn why using `Strings` for constant values is a really bad idea, and how enums give you greater flexibility and type safety, as well as making your code easier to write. From enums, you'll move into sealed classes, a particularly cool feature of Kotlin that lets you turbo-charge the concept of enums even further. You'll also dig into companion objects and factories, all of which contribute to a robust type-safe approach to programming where previously only `String` types were used.

Chapter 10: Functions and Functions and Functions   It may seem odd to have a chapter this late in the book that purports to focus on functions. However, as with most fundamentals in any discipline, you'll have to revisit the basics over and over again, shoring up weaknesses and adding nuance. This chapter does just that with functions. You'll dig more deeply into just how arguments really work, and how many options Kotlin provides to you in working with data going into and out of your functions.

Chapter 11: Speaking Idiomatic Kotlin   Kotlin, like all programming languages, has certain patterns of usage that seasoned programmers revert to time and time again. This chapter discusses these and some of the idioms of Kotlin. You'll get a jump start on writing Kotlin that looks like Kotlin is "supposed to" all while understanding how you have a tremendous amount of flexibility in choosing how to make your Kotlin programs feel like "you."

Chapter 12: Inheritance, One More Time, with Feeling   Yes, it really is another chapter on inheritance! This chapter takes what you've already learned about abstract classes and superclasses and adds interfaces and implementations into the mix. You'll also learn about the delegation pattern, a common Kotlin pattern that helps you take inheritance even further with greater flexibility than inheritance alone provides.

Chapter 13: Kotlin: The Next Step   No book can teach you everything you need to know, and this book is certainly no exception. There are some well-established places to look for next steps in your Kotlin programming journey, though, and this chapter gives you a number of jumping-off points to continue learning about specific areas of Kotlin.

Reader Support for This BookCompanion Download Files

As you work through the examples in this book, the project files you need are available for download from www.wiley.com/go/programmingkotlinapplications.

How to Contact the Publisher

If you believe you've found a mistake in this book, please bring it to our attention. At John Wiley & Sons, we understand how important it is to provide our customers with accurate content, but even with our best efforts an error may occur.

In order to submit your possible errata, please email it to our Customer Service Team at wileysupport@wiley.com with the subject line "Possible Book Errata Submission."

How to Contact the Author

We appreciate your input and questions about this book! Email me at brett@brettdmclaughlin.com, or DM me on Twitter at @bdmclaughlin.

1

Objects All the Way Down

WHAT'S IN THIS CHAPTER?

➤ A look at Kotlin syntax

➤ A brief history of Kotlin

➤ How Kotlin is like Java—and how it isn't

➤ Getting set up to code and run Kotlin

➤ Your first Kotlin program

➤ Why objects are cool (and why that matters)

KOTLIN: A NEW PROGRAMMING LANGUAGE

Kotlin, when you boil it all down, is just another programming language. If you're programming and writing code already, you'll pick up Kotlin quickly, because it has a lot in common with what you're already doing. That's even more the case if you're programming in an object-oriented language, and if you're coding in Java, well, Kotlin is going to feel very familiar, although different in some very nice ways.

If you're new to Kotlin, though, it's a great first language. It's very clear, it doesn't have lots of odd idioms (like, for example, Ruby or, god help us all, LISP), and it's well organized. You'll pick it up fast and find yourself comfortable quite quickly.

In fact, Kotlin is so straightforward that we're going to put aside a lot of explanation and history for now, and instead jump right into looking at some basic Kotlin code (check out Listing 1.1).

LISTING 1.1: A simple Kotlin program using classes and lists

```
data class User(val firstName: String, val lastName: String)

fun main() {
  val brian = User("Brian", "Truesby")
  val rose = User("Rose", "Bushnell")

  val attendees: MutableList<User> = mutableListOf(brian, rose)

  attendees.forEach {
    user -> println("$user is attending!")
  }
}
```

Take a minute or two to read through this code. Even if you've never looked at a line of Kotlin before, you can probably get a pretty good idea of what's going on. First, it defines a User class (and in fact, a special kind of class, a data class; more on that later). Then it defines a main function, which is pretty standard fare. Next up are two variables (or vals), each getting an instance of the User class defined earlier. Then a list is created called attendees and filled with the two users just created. Last up is a loop through the list, doing some printing for each.

If you ran this code, you'd get this rather unimpressive output:

```
User(firstName=Brian, lastName=Truesby) is attending!
User(firstName=Rose, lastName=Bushnell) is attending!
```

Obviously, parts of this probably look odd to you, whether you're brand new to writing code or an experienced Java pro. On top of that, it's likely you have no idea how to actually compile or run this code. That's OK, too. We'll get to all of that.

> **NOTE** *It bears repeating: You really don't need to understand the code in Listing 1.1. This book assumes you've programmed at least a little bit—and it's true that you'll likely understand Kotlin a bit faster if you have a Java background— but you will pick up everything you see in Listing 1.1 (and quite a bit more) just by continuing to read and working through the code samples. Just keep going, and you'll be programming in Kotlin in no time.*

For now, though, here's the point: Kotlin is really approachable, clean to read, and actually a pretty fun language to use. With that in mind, let's get some basics out of the way so you can get to writing code, not just looking at it.

WHAT IS KOTLIN?

Kotlin is an open-source programming language. It is, most notably, statically typed and object-oriented. Statically typed means that variables have types when you write your code and compile it, and those types are fixed. That also implies that Kotlin must be compiled, which is also true. Object-oriented means it has classes and inheritance, making it a familiar language for Java and C++ developers.

Kotlin was in fact created by a group of developers that worked on the JetBrains IDE, and it feels very much like a natural evolution of Java. It's been around in early form since 2011, but was officially released in 2016. That means it's new, which is good, but also means it's new, which at times can be bad. Kotlin is modern, can run inside a Java Virtual Machine (JVM), and can even be compiled to JavaScript—a cool feature we'll look at a little later.

It's also really important to note that Kotlin is a fantastic language for writing Android apps. In fact, many of its enhancements to Java reflect an Android usage. That said, even if you never intended to write a mobile app, you'll find Kotlin a welcome addition to your arsenal, and well suited for server-side programming.

What Does Kotlin Add to Java?

That's a good question that has a long answer. In fact, we'll spend most of this book answering that in various forms. But, for most, Kotlin adds or changes a few key features when compared to Java:

> **NOTE** *If you're new to Kotlin or not coming from a Java background, feel free to skip right on to the next section.*

- ➤ Kotlin ditches `NullPointerException` (and nullable variables altogether) in almost all situations.
- ➤ Kotlin supports extending functions without having to entirely override a parent class.
- ➤ Kotlin doesn't support checked exceptions (you may not find this to be an advancement, so fair warning).
- ➤ Kotlin adds components of functional programming, such as extensive lambda support and lazy evaluation.
- ➤ Kotlin defines data classes that let you skip writing basic getters and setters.

There's certainly a lot more to this list, but you can quickly see that Kotlin isn't just a slightly different version of Java. It seeks to be different and better, and in many ways, it very much is exactly that.

KOTLIN IS OBJECT-ORIENTED

At this point, most books and tutorials would have you put together a simple "Hello, World" program. That's all well and good, but the assumption here is that you want to get moving, and get moving quickly. For that reason, the logical place to begin with Kotlin is by creating an object.

An object is simple a programmatic representation of a thing. In the best case, that thing is a real-world object, like a car or a person or a product. For example, you could create an object to model a person like this:

```
class Person {

    /* This class literally does nothing! */

}
```

That's it. You can now create a new variable of type `Person` like this:

```
fun main() {
    val jennifer = Person()
}
```

If you put all this together into a single code listing, you'll have Listing 1.2.

LISTING 1.2: A very useless object in Kotlin (and a main function to use it)

```
class Person {

    /* This class literally does nothing! */

}

fun main() {
    val jennifer = Person()
}
```

Now, this is pretty lame code, honestly. It doesn't do anything, but it is object-oriented. Before we can improve it, though, you need to be able to run this good-for-almost-nothing code yourself.

INTERLUDE: SET UP YOUR KOTLIN ENVIRONMENT

Getting a Kotlin program to run is relatively easy, and if you're an old Java hand, it's actually *really* easy. You'll need to install a Java Virtual Machine and then a Java Development Kit (JDK). Then you'll want one of the numerous IDEs that support Kotlin. Let's take a blazing-fast gallop through that process.

Install Kotlin (and an IDE)

One of the easiest IDEs to use with Kotlin is IntelliJ IDEA, and starting with version 15, IntelliJ comes bundled with Kotlin. Plus, since IntelliJ is actually from JetBrains, you're getting an IDE built by the same folks who came up with Kotlin itself.

Install IntelliJ

You can download IntelliJ from www.jetbrains.com/idea/download. This page (shown in Figure 1.1) will then redirect you to the appropriate platform (Mac OS X for most of this book's examples). Download the free Community version to get started without any cost. Once the (rather large) download completes, install it (see Figure 1.2), and you'll get a Java Runtime Environment (JRE) and the JDK as part of installation.

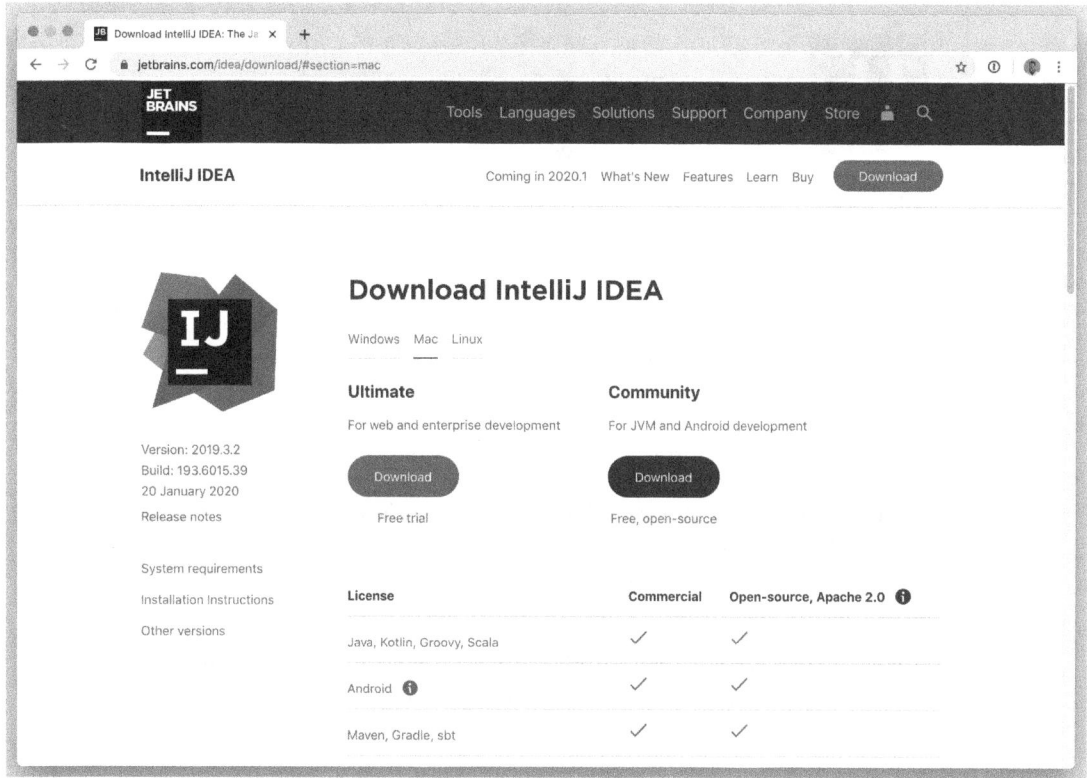

FIGURE 1.1: Download IntelliJ from the JetBrains download page.

> **NOTE** *IntelliJ is not the only IDE that works with Kotlin, and the list is actually growing pretty quickly. Other notable options are Android Studio (*developer .android.com/studio/preview/index.html*) and Eclipse (*www.eclipse.org/ downloads*). Eclipse in particular is immensely popular, but IntelliJ is still a great choice as it shares the JetBeans heritage with Kotlin.*

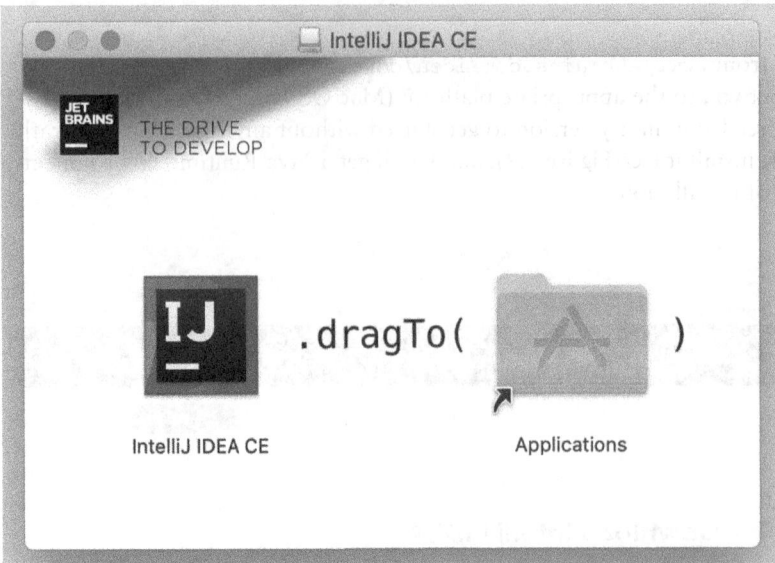

FIGURE 1.2: IntelliJ comes prepackaged with a system-specific installation process.

NOTE *The "installation process" for IntelliJ on Mac OS X is pretty simple: just drag the package (presented as an icon) into your Applications folder. You'll then need to go to that folder and launch IntelliJ or drag the icon into your Dock, which is what I've done.*

For Windows, you download the executable and run it. You can then create a shortcut on your desktop if you like.

In both cases, you can use the JetBrains Toolbox (which comes with the JetBrains Kotlin package) to keep your installation current and add updates when they're available.

You'll be given a pretty large number of options to get your IDE set up. For IntelliJ, you'll pick a UI theme (either is fine), a Launcher Script (I'd suggest you accept the default and let it create the script), the default plugins, and a set of featured plugins. You can click through these quickly, and then your IDE will restart. You'll see a welcome screen similar to Figure 1.3, and you should select Create New Project.

WARNING *You may need advanced permissions to install the Launcher Script that IntelliJ creates if you accepted the default location on Mac OS X.*

Be sure you select the Kotlin/JVM option when creating the project, as shown in Figure 1.4.

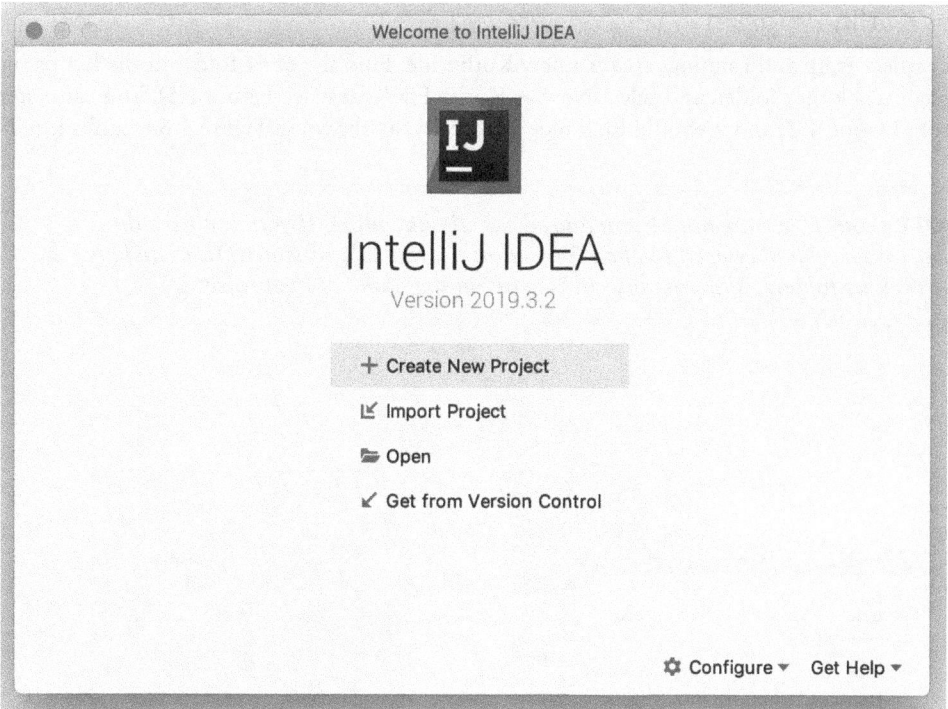

FIGURE 1.3: You'll generally be either creating a project from scratch or importing one from a code repository, like GitHub.

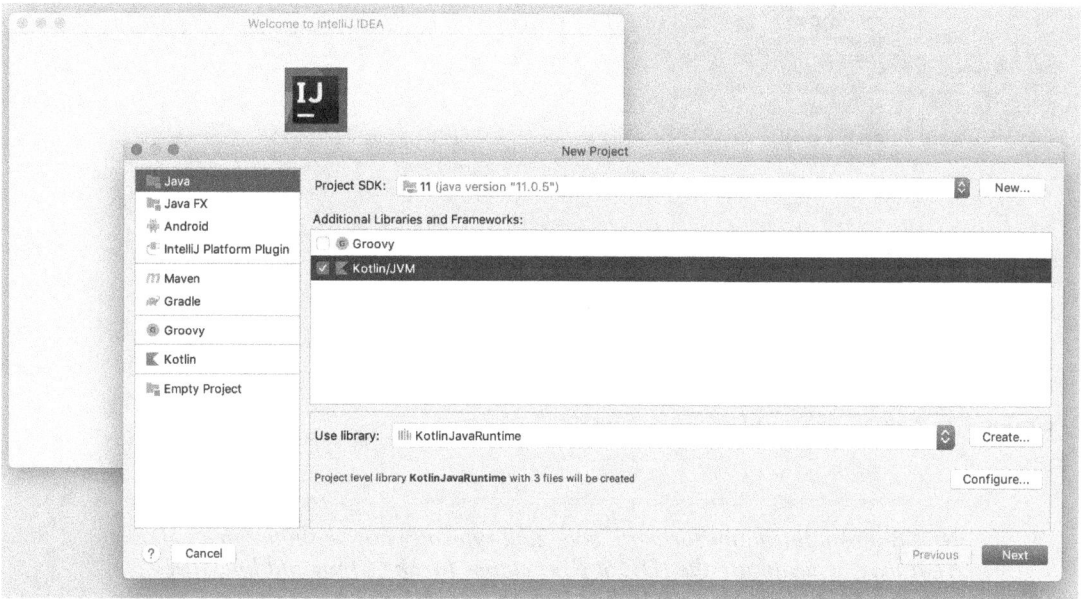

FIGURE 1.4: IntelliJ makes getting going in Kotlin simple and prompts you on creating a new project to include Kotlin libraries.

Create Your Kotlin Program

Once your project is up and running, create a new Kotlin file. Find the `src/` folder in the left navigation pane, right-click that folder, and select New ➤ Kotlin File/Class (see Figure 1.5). You can enter the code from Listing 1.2, and it should look nice and pretty, as shown in Figure 1.6 (thanks IntelliJ!).

> **NOTE** *Your IDE may not be configured exactly like mine. If you don't see the src/ folder, you may need to click Project on the left side of your IDE to display the various folders, and possibly click again on the name of the project.*

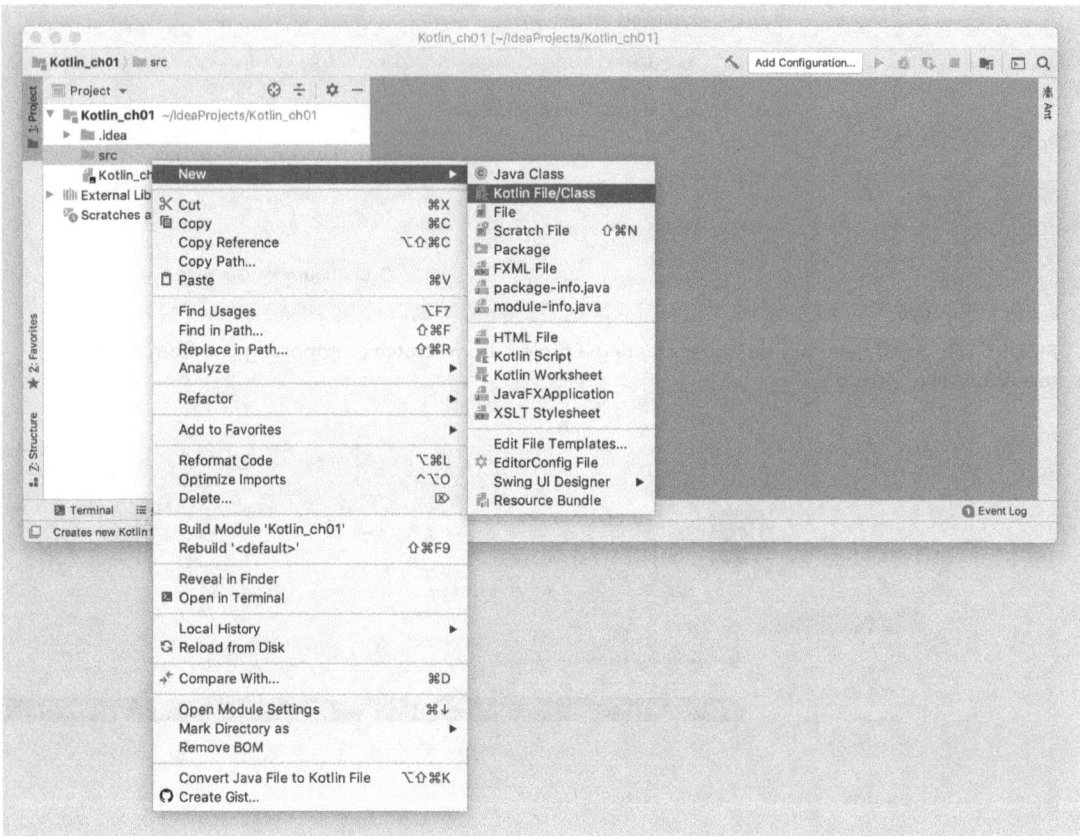

FIGURE 1.5: Kotlin code should go in the src/ folder.

> **NOTE** *From this point forward, code will typically not be shown in an IDE. That way, you can use the IDE of your choice (or the command line), because you should get the same results across IDEs.*

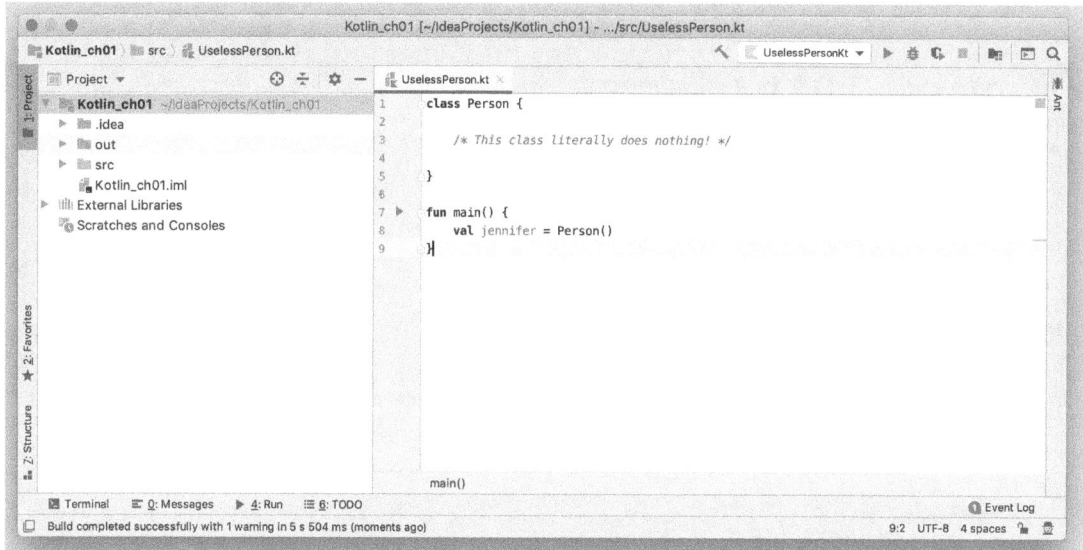

FIGURE 1.6: IntelliJ automatically formats code and adds sensible syntax highlighting.

Compile and Run Your Kotlin Program

All that's left now is to compile and run the program yourself. This is easy, because IntelliJ gives you a convenient little green arrow to click when you have a Kotlin file with a main() function defined. Just hover over the arrow and click (see Figure 1.7). You can then select Run and your filename (I named mine "UselessPerson"). Your program will be compiled and run, with the output shown in a new pane at the bottom of the IDE (see Figure 1.8).

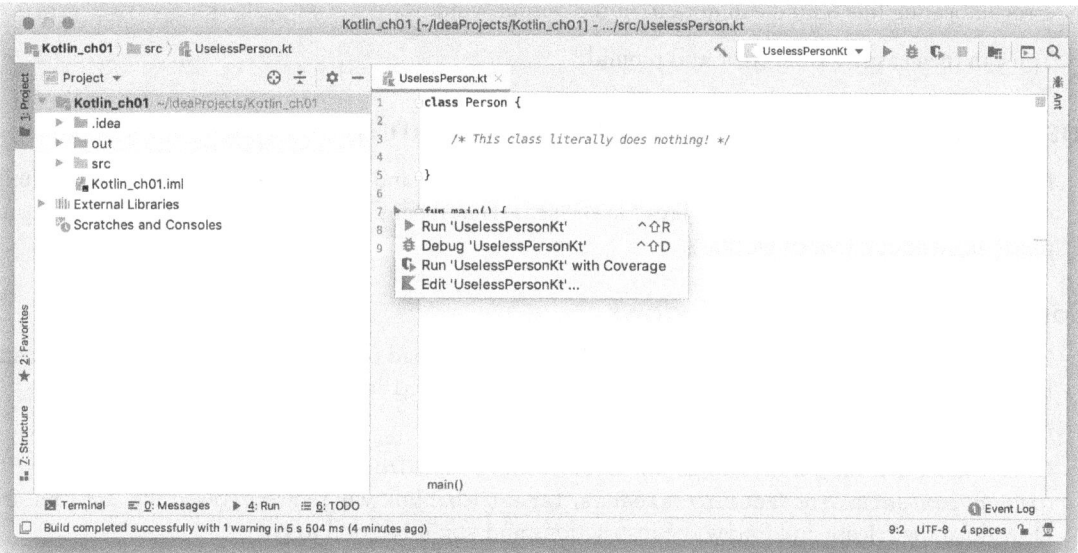

FIGURE 1.7: You can click the green Run button and select the first option to build and run your code.

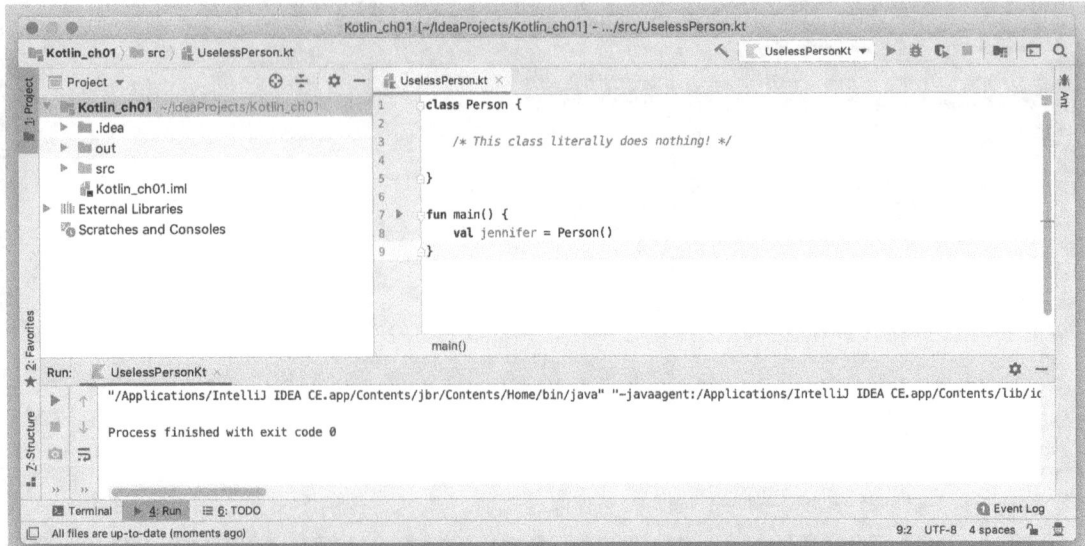

FIGURE 1.8: The empty output of your program (which will soon be non-empty) displays in its own window.

In this case, you shouldn't get any errors, but there's not any output either. We'll fix that shortly.

Fix Any Errors as They Appear

One last note before getting back to improving that useless `Person` class. IntelliJ and all other IDEs are great at giving you visual indicators when there is a problem with your code. For example, Figure 1.9 shows IntelliJ once it's tried to compile the same program with an error. In this case, the open and close parentheses are missing from line 8. You'll see an orange indicator in the code editor and an error indicating line 8 (and column 20) in the output window.

You can then easily fix the error and rebuild.

Install Kotlin (and Use the Command Line)

For power users, there's a tendency to want to use the command line for nearly everything. Kotlin is no exception. Because it's "mostly Java" in the sense that it runs using a JVM and JDK, you can get pretty far without a lot of work.

Command-Line Kotlin on Windows

For Windows users, you'll first need a JDK. You can download one from the Oracle Java download page at www.oracle.com/technetwork/java/javase/downloads. That download has version-specific instructions that are easy to follow.

Once you have Java, you need to get the latest Kotlin release from GitHub. You can find that at github.com/JetBrains/kotlin/releases/latest (that link will redirect you to the latest release). Download the release and follow the instructions and you'll be good to go.

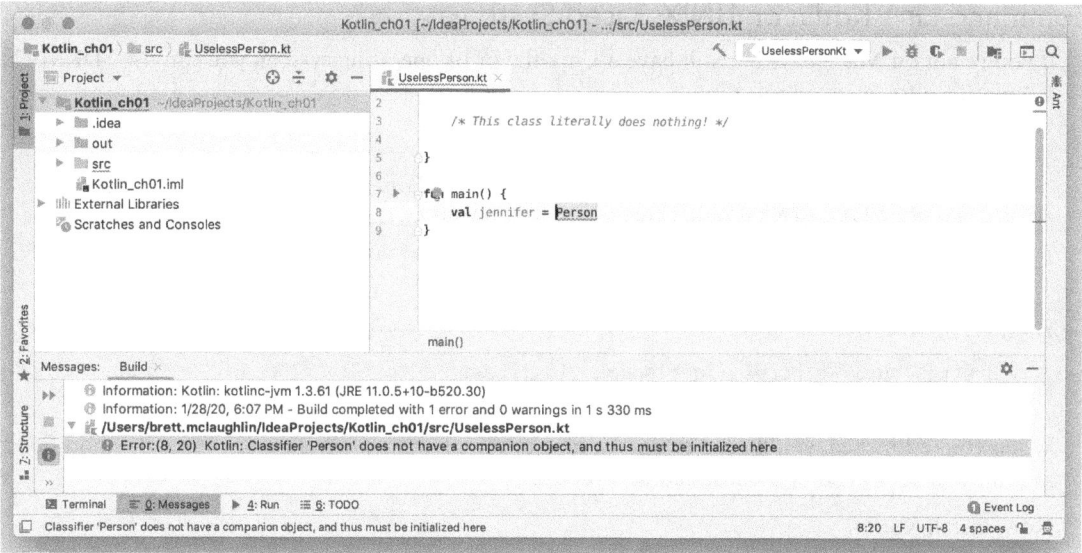

FIGURE 1.9: Good IDEs help you quickly find and fix errors.

> **NOTE** *These instructions are intentionally a bit sparse. If you're using the command line already, you probably don't need a lot of hand holding. For almost everyone else, though, using an IDE really is the best approach. As a bonus, you can also use IntelliJ as a proxy for the compiler, so you may just want to save the time it would take you to mess with the command line and put it into coding Kotlin!*

Command-Line Kotlin on Mac OS X

The easiest path to getting Kotlin working on Mac OS X is to use one of the package managers popular on Macs: either Homebrew (brew.sh) or MacPorts (www.macports.org). Both of these make getting Kotlin up and running trivial.

For MacPorts, just run the following command:

```
brett $ sudo port install kotlin
```

This requires elevated permissions, but after it runs, you'll be all set.

For Homebrew, first do an update:

```
brett $ brew update
```

Next up, install Kotlin:

```
brett $ brew install kotlin
```

Command-Line Kotlin on UNIX-Based Systems

If you're not on Mac OS X but still have a Unix flavor of operating system, you can use SDKMAN! (sdkman.io) for installing Kotlin.

> **NOTE** *To be accurate, Mac OS X is a Unix-based operating system, so you can use the SDKMAN! instructions for Macs instead of Homebrew or MacPorts.*

First, get SDKMAN!:

```
brett $ curl -s https://get.sdkman.io | bash
```

When you're finished, you'll need to open a new terminal or shell window or source the modified file as indicated at the end of the installation process.

Now, install Kotlin:

```
brett $ sdk install kotlin
```

Verify Your Command-Line Installation

However you've chosen to install Kotlin, when you're finished, you should be able to validate your installation with this command:

```
brett $ kotlinc
```

> **WARNING** *At this point, you may get prompted to install a Java runtime. This should fire up your system to handle this, and you can accept the prompts without a lot of worry. Find the JDK or JRE for your system, download it, and run it. Then come back and try out* kotlinc *again.*

If you have your system appropriately configured with Java, you should get back something like this:

```
brett $ kotlinc

Java HotSpot(TM) 64-Bit Server VM warning: Options -Xverify:none and -noverify were
deprecated in JDK 13 and will likely be removed in a future release.

Welcome to Kotlin version 1.3.61 (JRE 13.0.2+8)

Type :help for help, :quit for quit

>>>
```

This is the Kotlin REPL (Read-Eval-Print Loop), a tool for quickly evaluating Kotlin statements. We'll look at this in more detail later, but for now, exit the REPL by typing **:quit**.

You can also verify your version of Kotlin with the following command:

```
brett $ kotlin -version

Kotlin version 1.3.61-release-180 (JRE 13.0.2+8)
```

At this point, you're ready to roll!

CREATING USEFUL OBJECTS

With a working Kotlin environment, it's time to go make that `Person` class from Listing 1.2 something less vacuous. As mentioned earlier, objects should model real-world objects. Specifically, if an object represents a "thing" in the world (or in formal circles you'll sometimes hear that objects are "nouns"), then it should have the properties or attributes of that thing, too.

A person's most fundamental property is their name, specifically a first name and last name (or surname, if you like). These can be represented as properties of the object. Additionally, these are required properties; you really don't want to create a person without a first and last name.

For required properties, it's best to require those properties when creating a new instance of an object. An *instance* is just a specific version of the object; so you might have multiple instances of the `Person` class, each one representing a different actual person.

> **WARNING** *As you may already be figuring out, there's a lot of technical vocabulary associated with classes. Unfortunately, it will get worse before it gets better: instances and instantiation and constructors and more. Don't get too worried about catching everything right away; just keep going, and you'll find that the vocabulary becomes second nature faster than you think. In Chapter 3, you'll dive in deeper, and in fact, you'll keep revisiting and growing your class knowledge throughout the entire book.*

Pass In Values to an Object Using Its Constructor

An object in Kotlin can have one or more constructors. A *constructor* does just what it sounds like: it constructs the object. More specifically, it's a special method that runs when an object is created. It can also take in property values—like that required first and last name.

You put the constructor right after the class definition, like this:

```
class Person constructor(firstName: String, lastName: String) {
```

In this case, the constructor takes in two properties: `firstName` and `lastName`, both `String` types. Listing 1.3 shows the entire program in context, along with creating the `Person` instance by passing in the values for `firstName` and `lastName`.

> **WARNING** *You'll sometimes hear properties or property values called parameters. That's not wrong; a parameter is usually something passed to something else; in this case, something (a first name and last name) passed to something else (a constructor). But once they're assigned to the object instance, they're no longer parameters. At that point, they are properties (or more accurately, property values) of the object. So it's just easier to call that a property value from the start.*

See? The terminology is confusing. Again, though, it will come with time. Just keep going.

LISTING 1.3: A less useless object in Kotlin and its constructor

```kotlin
class Person constructor(firstName: String, lastName: String) {

    /* This class still doesn't do much! */

}

fun main() {
    val brian = Person("Brian", "Truesby")
}
```

Now the class takes in a few useful properties. But, as most developers know, there's a tendency to condense things in code. There's a general favoring of typing less, rather than typing more. (Note that this rarely applies to book authors!) So Listing 1.3 can be condensed; you can just drop the word `constructor` and things work the same way. Listing 1.4 shows this minor condensation (and arguable improvement).

LISTING 1.4: Cutting out the constructor keyword

```kotlin
class Person(firstName: String, lastName: String) {

    /* This class still doesn't do much! */

}

fun main() {
    val brian = Person("Brian", "Truesby")
}
```

Print an Object with toString()

This is definitely getting a little better. But the output is still empty, and the class is still basically useless. However, Kotlin gives you some things for free: most notably for now, every class automatically gets a `toString()` method. You can run this method by creating an instance of the class (which you've already done) and then calling that method, like this:

```
val brian = Person("Brian", "Truesby")

println(brian.toString())
```

Make this change to your `main` function. Create a new `Person` (give it any name you want), and then print the object instance using `println` and passing into `println` the result of `toString()`.

> **NOTE** *You may be wondering where in the world that* `toString()` *method came from. (If not, that's OK, too.) It does seem to sort of magically appear. But it's not magical it all. It's actually inherited. Inheritance is closely related to objects, and something we'll talk about in a lot more detail in both Chapter 3 and Chapter 5.*

Terminology Update: Functions and Methods

A few specifics on terminology again. A *function* is a piece of code that runs. `main` is an example of a function. A *method* is, basically, a function that's attached to an object. Put another way, a method is also a piece of code, but it's not "stranded" and self-standing, like a function is. A method is defined on an object and can be run against a specific object instance.

> **NOTE** *In much of the official Kotlin documentation, there is not a clear distinction drawn between a function and a method. However, I've chosen to draw this distinction because it's important in general object-oriented programming, and if you work or have familiarity with any other object-based language, you'll run across these terms. But you should realize that in "proper" Kotlin, all methods are functions.*

That last sentence is important, so you may want to read it again. It's important because it *means* that a method can interact with the object instance. For example, a method might want to use the object instance's property values, like, say, a first name and last name. And with that in mind, back to your code!

Print an Object (and Do It with Shorthand)

You can run the `println` function at any time, and you just pass it something to print. So you could say:

```
println("How are you?")
```

and you'd just get that output in your results window. You can also have it print the result from a method, like `toString()`, which is what you did earlier. But there's another shortcut. If you pass in something to `println()` that has a `toString()` method, that method is automatically run. So you can actually trim this code:

```
println(brian.toString())
```

down to just this:

```
println(brian)
```

In the latter case, Kotlin sees an object passed to `println()` and automatically runs `brian.toString()` and passes the result on for printing. In either case, you'll get output that looks something like this:

```
Person@7c30a502
```

That's not very useful, is it? It's essentially an identifier for your specific instance of `Person` that is useful to Kotlin internals and the JVM, but not much else. Let's fix that.

Override the toString() Method

One of the cool things about a class method is that you can write code and define what that method does. We haven't done that yet, but it's coming soon. In the meantime, though, what we have here is slightly different: a method that we *didn't* write code for, and that *doesn't* do what we want.

In this case, you can do something called *overriding* a method. This just means replacing the code of the method with your own code. That's exactly what we want to do here.

First, you need to tell Kotlin that you're overriding a method by using the `override` keyword. Then you use another keyword, `fun`, and then the name of the method to override, like this:

```
override fun toString()
```

> **NOTE** *Earlier, you learned the difference between a function and a method. And* `toString()` *is definitely a method, in this case on* `Person`. *So why are you using the* `fun` *keyword? That looks an awful lot like "function," and that's, in fact, what it stands for.*
>
> *The official answer is that Kotlin essentially sees a method as a function attached to an object. And it was easier to not use a different keyword for a standalone function and an actual method.*
>
> *But, if that bugs you, you're in good company. It bugs me, too! Still, for the purposes of Kotlin, you define both functions and methods with* `fun`.

But `toString()` adds a new wrinkle: it returns a value. It returns a `String` to print. And you need to tell Kotlin that this method returns something. You do that with a colon after the parentheses and then the return type, which in this case is a `String`:

```
override fun toString(): String
```

Now you can write code for the method, between curly braces, like this:

```kotlin
class Person(firstName: String, lastName: String) {

    override fun toString(): String {
        return "$firstName $lastName"
    }
}
```

This looks good, and you've probably already figured out that putting a dollar sign ($) before a variable name lets you access that variable. So this takes the firstName and lastName variables passed into the Person constructor and prints them, right?

Well, not exactly. If you run this code, you'll actually get the errors shown in Figure 1.10.

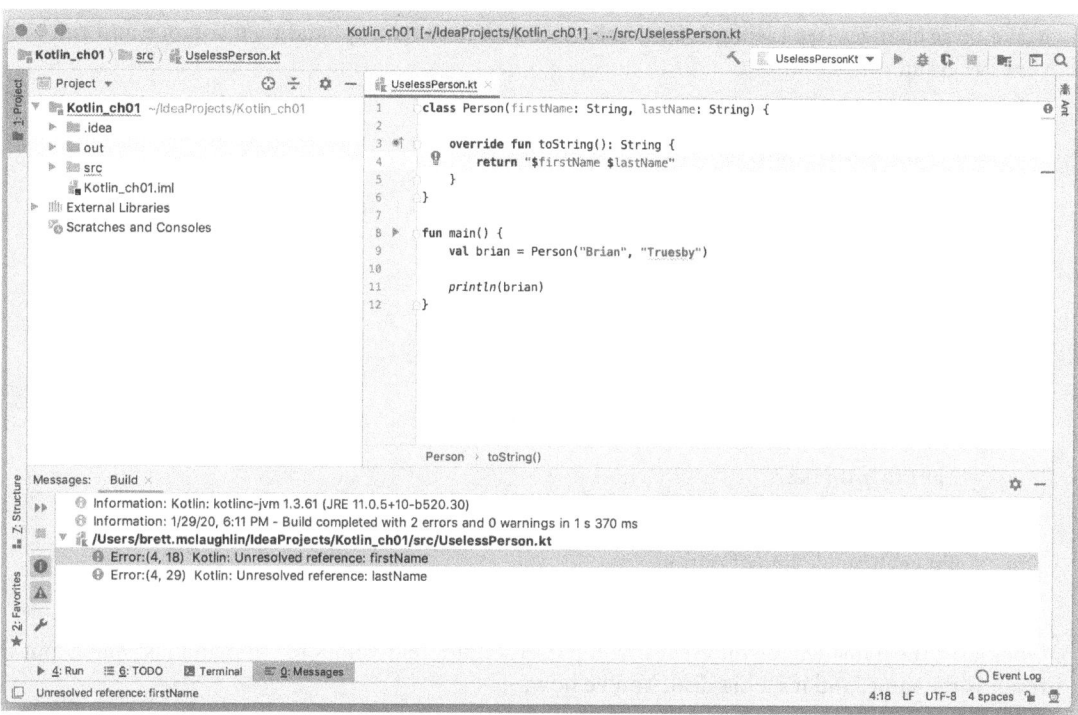

FIGURE 1.10: Why doesn't this override of toString() work?

What gives here? Well, it turns out to be a little tricky.

All Data Is Not a Property Value

You have a constructor and it takes in two pieces of data: firstName and lastName. That's controlled by the constructor declaration:

```kotlin
class Person(firstName: String, lastName: String) {
```

But here's what is tricky: just accepting those values does not actually turn them into property values. That's why you get the error in Figure 1.10; your `Person` object accepted a first and last name, but then promptly ignored them. They aren't available to be used in your `toString()` overridden method.

You need to use the `val` keyword in front of each piece of data to turn that data into property values. Here's the change you need to make:

```
class Person(val firstName: String, val lastName: String) {
```

Specifically, by using `val` (or `var`, which we'll talk about shortly), you've created variables, and assigned them to the `Person` instance being created. That then allows those variables (or properties, to be even more precise) to be accessed, like in your `toString()` method.

Make these changes (see Listing 1.5 to make sure you're caught up) and then compile and run your program.

LISTING 1.5: Converting data to actual properties

```
class Person(val firstName: String, val lastName: String) {

    override fun toString(): String {
        return "$firstName $lastName"
    }
}

fun main() {
    val brian = Person("Brian", "Truesby")

    println(brian)
}
```

You should get a single line of output:

```
Brian Truesby
```

Obviously, the name will be different if you passed in different values for first and last name, but the result is the same, and it's a big deal. You've now:

➤   Created a new object

➤   Defined a constructor for the object

➤   Accepted two pieces of data in that constructor and stored them as properties associated with the object instance

➤   Overridden a method and made it useful

➤   Written a `main` function

➤   Instantiated your custom object and passed in values

➤   Used the object to print itself out, using your overridden method

Not too bad for getting started! There's just one more detail to work through before closing shop on your first foray into Kotlin.

INITIALIZE AN OBJECT AND CHANGE A VARIABLE

Suppose you want to play around a bit with your `Person` class. Try this out: update your code to match Listing 1.6 (some of this may be confusing, but you can probably figure out most of what's going on).

LISTING 1.6: Creating a new property for a Person

```kotlin
class Person(val firstName: String, val lastName: String) {
    val fullName: String

    // Set the full name when creating an instance
    init {
        fullName = "$firstName $lastName"
    }

    override fun toString(): String {
        return fullName
    }
}

fun main() {
    // Create a new person
    val brian = Person("Brian", "Truesby")

    // Create another person
    val rose = Person("Rose", "Bushnell")

    println(brian)
}
```

You'll see a number of new things here, but none are too surprising. First, a new variable is declared inside the `Person` object: `fullName`. This is something you've already done in your `main` function. This time, though, because you're doing it inside the `Person` object, it automatically becomes part of each `Person` instance.

Another small change is the addition of a new `Person` instance in `main`; this time it's a variable named `rose`.

Then, there's a new keyword: `init`. That bears further discussion.

Initialize a Class with a Block

In most programming languages, Java included, a constructor takes in values (as it does in your `Person` class) and perhaps does some basic logic. Kotlin does this a bit differently; it introduces the idea of an initializer block. It's this block—identified conveniently with the `init` keyword—where you put code that should run every time an object is created.

This is a bit different than you may be used to: data comes in through a constructor, but it's separated from the initialization code, which is in the initializer block.

In this case, the initializer block uses the new `fullName` variable and sets it using the first and last name properties passed in through the class constructor:

```
// Set the full name when creating an instance
init {
    fullName = "$firstName $lastName"
}
```

Then this new variable is used in toString():

```
    override fun toString(): String {
    return fullName
}
```

> **WARNING** *As much new material as this chapter introduces, you may have just run across the most important thing that you may learn, in the long-term sense. By changing* `toString()` *to use* `fullName`, *rather than also using the* `firstName` *and* `lastName` *variables directly, you are implementing a principle called DRY: Don't Repeat Yourself.*
>
> *In this case, you're not repeating the combination of first name and last name, which was done already in the initializer block. You assign that combination to a variable, and then forever more, you should use that variable instead of what it actually references. More on this later, but take note here: this is a big deal!*

Kotlin Auto-Generates Getters and Setters

At this point, things are going well. Part of that is all you've added, but another big help is that Kotlin is doing a lot behind the scenes. It's running code for you automatically (like that initializer block) and letting you override methods.

It's doing something else, too: it's auto-generating some extra methods on your class. Because you made `firstName` and `lastName` property values (with that `val` keyword), and you defined a `fullName` property, Kotlin created getters and setters for all of those properties.

Terminology Update: Getters, Setters, Mutators, Accessors

A getter is a method that allows you to get a value. For instance, you can add this into your `main` function, and it will not only work, but print out just the first name of the `brian` `Person` instance:

```
// Create a new person
val brian = Person("Brian", "Truesby")
println(brian.firstName)
```

This works because you have a getter on `Person` for `firstName`. You can do the same with `fullName` and `lastName`, too. This getter is, more formally, an *accessor*. It provides access to a value, in this case a property of `Person`. And it's "free" because Kotlin creates this accessor for you.

Kotlin also gives you a setter, or (again, more formally) a *mutator*. A mutator lets you mutate a value, which just means to change it. So you can add this into your program:

```
// Create a new person
val brian = Person("Brian", "Truesby")
println(brian.firstName)

// Create another person
val rose = Person("Rose", "Bushnell")
rose.lastName = "Bushnell-Truesby"
```

Just as you can get data through an accessor, you can update data through mutators.

> **WARNING** *For the most part, I'll be calling getters accessors, and calling setters mutators. That's not as common as "getter" or "setter," but as a good friend and editor of mine once told me, a setter is a hairy and somewhat fluffy dog; a mutator lets you update class data. The difference—and his colorful explanation—has stuck with me for 20 years.*

Now, if you've gone ahead and compiled this code, you've run into yet another odd error, and that's the last thing to fix before moving on from this initial foray into objects.

Constants Can't Change (Sort of)

Here's the code causing the problem:

```
// Create another person
val rose = Person("Rose", "Bushnell")
rose.lastName = "Bushnell-Truesby"
```

If you try to run this code, you'll get an error like this:

```
Error: Kotlin: Val cannot be reassigned
```

One of the things that is fairly unique about Kotlin is its strong stance on variables. Specifically, Kotlin allows you to not just declare the type of a variable, but also whether that variable is a *mutable* variable, or a *constant* variable.

> **NOTE** *The terminology here is a bit confusing, so take your time. Just as with methods being declared with the* fun *keyword, the idea of a constant variable takes a little getting used to.*

When you declare a variable in Kotlin, you can use the keyword `val`, as you've already done:

```
val brian = Person("Brian", "Truesby")
```

But you can also use the keyword var, something you *haven't* done yet. That would look like this:

```
var brian = Person("Brian", "Truesby")
```

First, in both cases, you end up with a variable; val does not stand for value, for example, but is simply another way to declare a variable, alongside var. When you use val, you are creating a constant variable. In Kotlin, a constant variable can be assigned a value once, and only once. That variable is then constant and can never be changed.

You created the lastName variable in Person with this line:

```
class Person(val firstName: String, val lastName: String) {
```

That defines lastName (and firstName) as a constant variable. Once it's passed in and assigned when the Person instance is created, it can't be changed. That makes this statement illegal:

```
rose.lastName = "Bushnell-Truesby"
```

To clear up the odd error from earlier, what you need instead is for lastName to be a *mutable* variable; you need it to be changeable after initial assignment.

> **NOTE** *Not to beat a hairy and somewhat fluffy dog to death, but here is another reason to use mutator over setter; a mutator allows you to mutate a mutable variable. This aligns the terminology much more cleanly than using "setter."*

So change your Person constructor to use var instead of val. This indicates that firstName and lastName can be changed:

```
class Person(var firstName: String, var lastName: String) {
```

Now you should be able to compile the program again, without error. In fact, once you've done that, make a few other tweaks. You want to end up with your code looking like Listing 1.7.

LISTING 1.7: Using mutable variables

```kotlin
class Person(var firstName: String, var lastName: String) {
    var fullName: String

    // Set the full name when creating an instance
    init {
        fullName = "$firstName $lastName"
    }

    override fun toString(): String {
        return fullName
    }
}
```

```kotlin
fun main() {
    // Create a new person
    val brian = Person("Brian", "Truesby")
    println(brian)

    // Create another person
    val rose = Person("Rose", "Bushnell")
    println(rose)

    // Change Rose's last name
    rose.lastName = "Bushnell-Truesby"
    println(rose)
}
```

Here, fullName has been made mutable, and there's a little more printing in main. It should compile and run without error now.

But wait! Did you see your output? There's a problem! Here's what you probably get when you run your code:

```
Brian Truesby
Rose Bushnell
Rose Bushnell
```

Despite what Meat Loaf had to say about two out of three not being bad, this is not great. Why is Rose's name in the last instance not printing with her new last name?

Well, to solve that, it's going to take another chapter, and looking a lot more closely at how Kotlin handles data, types, and more of that automatically running code.

2

It's Hard to Break Kotlin

WHAT'S IN THIS CHAPTER?

➤ Dive deeper into Kotlin types and variables

➤ More on mutability

➤ Creating custom accessors and mutators

➤ Assignment, during and after variable creation

➤ Type safety and what it means

UPGRADE YOUR KOTLIN CLASS GAME

You've already had a solid introduction to Kotlin, including building your first object. So we're going to get right to work. There's that nagging issue with a person's last name showing incorrectly, but before fixing that, you're ready to up how you build classes from a pre-compile point of view.

So far, you've used a single file to contain both your class (Person) and your main function. That's not scalable, though. Before long, you're going to have lots of classes—and that's classes, not instances of classes.

> **NOTE** *As a quick refresher, a class is a definition. You have a* Person *class, and you might also have a* Car *class or a* House *class. An instance of a class is a specific piece of memory that holds a representation of an object. In the last chapter, when you created a* brian *or a* rose *variable, you assigned those instances of the* Person *class.*
>
> *It's pretty easy to get lots of instances of a class, and then to also have lots of different classes defined as well. In those cases, things get messy if you're just using a single file.*

If you throw all your classes into one file, that file is going to be a mess. It's better to separate each of your classes into their own files: one file per class (in general), and then if you have a `main` function, give that its own file, too.

Name a File According to Its Class

Go ahead and create a new Kotlin project. You can call it whatever you like; `Kotlin_ch02` works if you need a suggestion. Then, in the `src/` folder, create a new file called `Person`. Your IDE will likely add the `.kt` extension for you. Then, just drop in your existing `Person` code from Chapter 1. That code is shown again in Listing 2.1 for good measure.

LISTING 2.1: The Person code from Chapter 1

```kotlin
class Person(var firstName: String, var lastName: String) {
    var fullName: String

    // Set the full name when creating an instance
    init {
        fullName = "$firstName $lastName"
    }

    override fun toString(): String {
        return fullName
    }
}
```

Now create another file, also in `src/`, and call it `PersonApp`. You can then drop your existing `main` function into that. That bit of code is reprised in Listing 2.2.

LISTING 2.2: Main function for testing the Person class

```kotlin
fun main() {
    // Create a new person
    val brian = Person("Brian", "Truesby")
    println(brian)

    // Create another person
    val rose = Person("Rose", "Bushnell")
    println(rose)

    // Change Rose's last name
    rose.lastName = "Bushnell-Truesby"
    println(rose)
}
```

At this point, you should be able to run `PersonApp` just as you did in Chapter 1. Your output should also be the same. You can see in Figure 2.1 the IntelliJ IDE, with the two Kotlin files as tabs, and the output in the bottom pane.

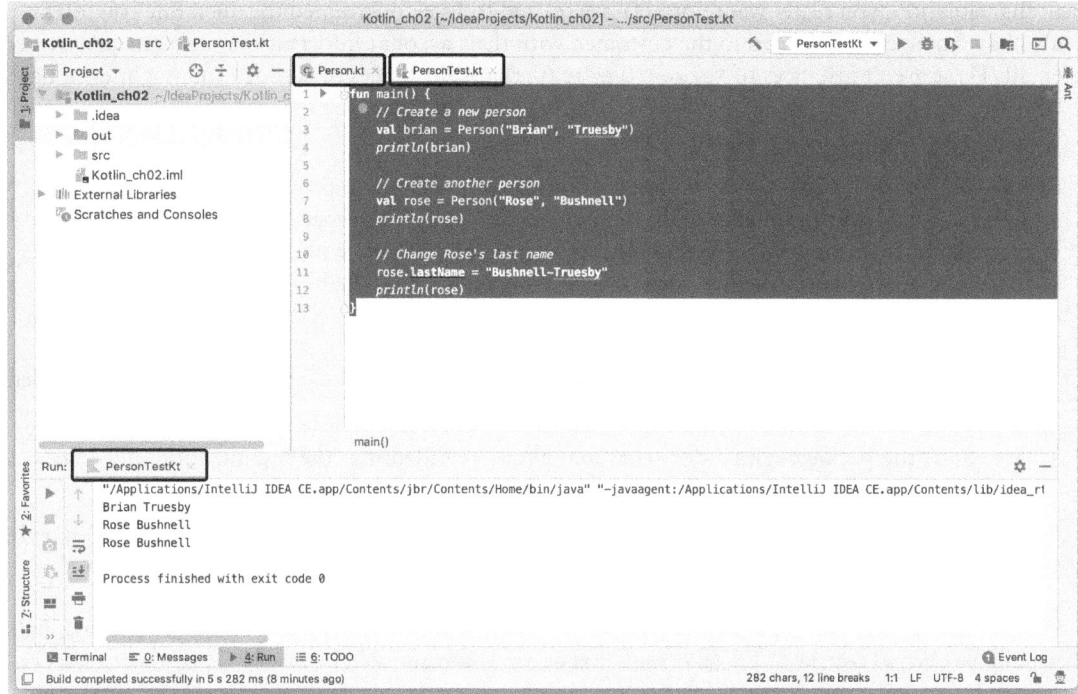

FIGURE 2.1: Your IDE should let you easily get to your source code as well as see output.

> **WARNING** *If you get any warnings, the most likely culprit is that you didn't create your* Person *and* PersonApp *files in the same directory. Make sure both files are in the* src/ *directory and try again.*

At this point, you could go ahead and keep creating more objects in the src/ directory. But there's still a better way!

Organize Your Classes with Packages

While it's great that you've got Person separate from the test main function, you've still not really solved the original problem: class organization. If having all your classes in one file is a pain, then by extension, so is having all your classes in a single directory.

What you really want is to organize your classes by functionality. So let's assume that you want your Person class, but you also think you might later want some specific types of people; maybe a Mother class or a Child class. Suppose you're building software to model a genealogy, for example.

In that situation, you really want your classes related to `Person` grouped together. But you might also have some classes related to the customer, with their account information: username, password, email. These aren't instances of `Person`, they're (perhaps) instances of a `User`. You might even have a `UserPreferences` class at some point, too, letting you store that user's information. So that might all belong to a group of classes related to `User`.

Kotlin—and many other languages—use the concept of packages to provide this organization. A *package* is just a logical grouping of classes, with a particular namespace: some way to prefix and identify classes as being related. You could have a `person` package with `Person`, `Mother`, `Father`, and so forth, and a `user` package with `User`, `UserPreferences`, and the like.

Naming your package is important, though, so a few best practices:

➤   Packages should begin with lowercase letters. Use `person` instead of `Person`. (Classes begin with capitals; packages with lowercase.)

➤   Start the package with a dot-separated name that identifies the organization, like `org.wiley` or `com.myCodeShop`.

➤   If you might use languages other than Kotlin, you may want to also add `kotlin` as part of the package name, to distinguish the package from other code (such as Java).

Put all this together and separate the pieces with a dot delimiter (.). For this book, the package will usually be `org.wiley.kotlin.[general category]`. `Person` will be in `org.wiley.kotlin.person` and `User` will be in `org.wiley.kotlin.user`.

> **NOTE** *You don't always have to have the last part of the package name (like* person) *match a class in that package (like* Person). *You might have an* org.wiley.kotlin.sport *package that has various sports-related classes like* Football, Baseball, Hockey, *and* Volleyball, *but not have an actual* Sport *class. It really depends on your own preferences and style.*

Put Person in a Package

With a package decided, you need to actually create this structure. Again, this is where an IDE can really make things simpler. If you're using IntelliJ, right-click your `src/` folder and choose New and then Package. Type in the package name (`org.kotlin.wiley.person` if you're following along, or use your own name), and you'll see a small package indicator created as part of your `src/` tree.

> **WARNING** *Each IDE handles this a bit differently, but the similarities are far greater than the differences. You can always just play around with the IDE to look for a New Package option (that's what I did when I was first learning Kotlin in IntelliJ), or check your IDE's documentation.*
>
> *If you're using the command line exclusively, it's a bit more of a pain. You'll need to create a directory structure to match your package structure, so you'd need an* org/ *directory, and then* wiley/, *then* kotlin/, *then* person/, *and then put your* Person *class there. You'll also need to work with the* kotlinc *compiler a bit more.*
>
> *You can find this documentation easily online, so to keep things moving, most of this book will assume you're either using an IDE or comfortable looking up how to compile files in packages on your own.*

Now comes the "thank goodness for an IDE" part: just click your Person class in the left pane and drag it into the new package indicator. You'll get a dialog box that looks like Figure 2.2.

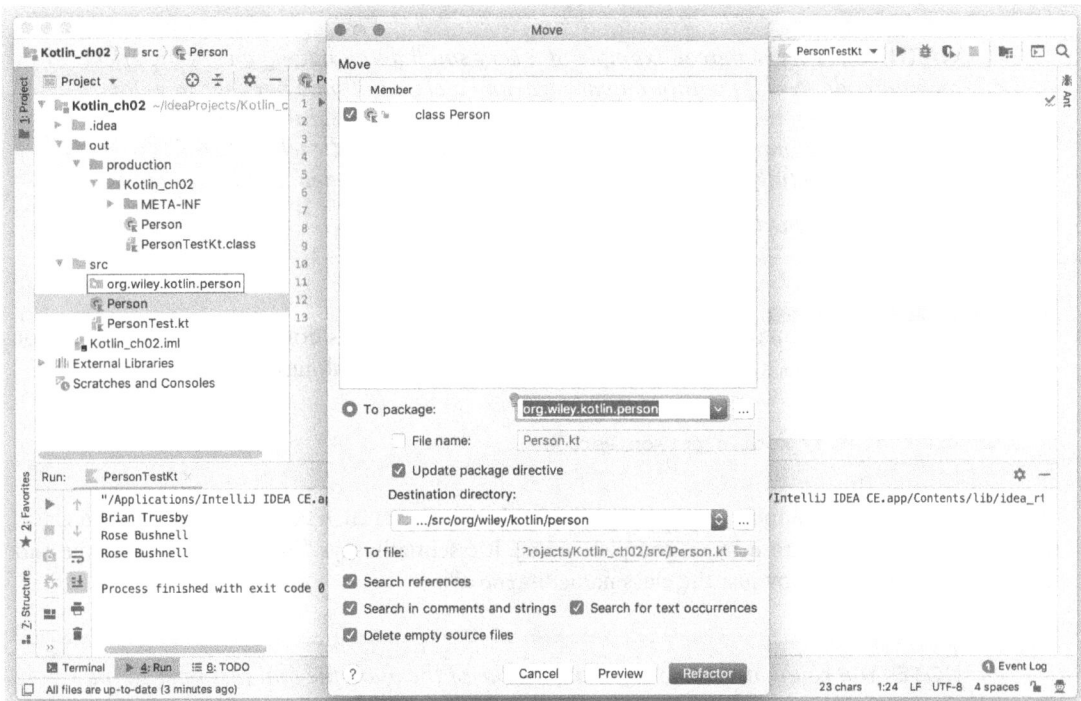

FIGURE 2.2: Moving a class to a new package is basically a refactoring, which your IDE can handle.

This dialog will ensure that yes, you really do want to move the class into the indicated package. You can leave all these defaults, as IntelliJ is happy to do this work for you. When you're ready, click the Refactor button.

Now, your navigation pane (on the left in the IDE) should show `Person` under the new `org.wiley.kotlin.person` package, as shown in Figure 2.3.

Before you think that too much magic has gone on, take a close look at the first line of your `Person` class, which is new:

```
package org.wiley.kotlin.person
```

That line tells Kotlin that the class belongs to the `org.kotlin.wiley.person` package. The compiled class then *actually* is `org.kotlin.wiley.person.Person`, not just `Person`.

This is going to cause your `main` function to break, because that function doesn't know anything about `org.kotlin.wiley.person.Person`; and now there's no `Person` class.

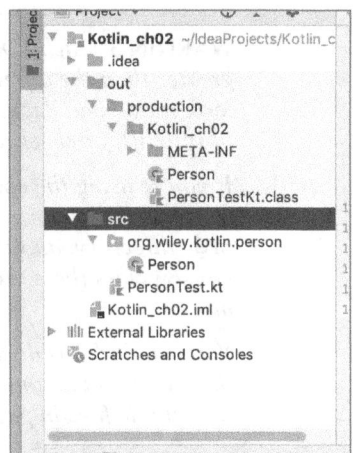

FIGURE 2.3: Person is now nested under its containing package.

> **WARNING** *This is a great example of a very small detail having a very big effect on your code. A class without a package, and a class within a package, are not the same at all—even if they are named the same. You could theoretically have multiple classes named the same in* multiple *packages, and each is distinct from the other. Hopefully you're already thinking this is a bad idea; but it is possible.*
>
> *The point is important, though: naming matters.*

Normally, you'd need to make a change here, but again, the IDE has done some nice work for you. Check out your file with `main` in it (`PersonApp` if you've followed along); it also has a new line at the top:

```
import org.wiley.kotlin.person.Person
```

This line tells Kotlin to import, or make available, the `org.kotlin.wiley.person.Person` class *without* needing to refer to it by its complete name. It essentially says "import that long name and allow it to be referenced by just the class name: `Person`."

> **NOTE** *You could also leave out that top line of the main file and just refer to the class by its entire name every time you use it. So you'd create a new instance like this:*
>
> ```
> val brian = org.wiley.kotlin.person.Person("Brian", "Truesby")
> ```
>
> You'd have to do this for every reference to `Person`, though. Developers don't typically like this, as it's a lot of extra typing, so using `import` is a lot more common.

With all these changes in place, you can recompile and run your program again. Things should work beautifully (although that last name is still wrong!).

Classes: The Ultimate Type in Kotlin

Before getting further into Kotlin types—something this chapter is going to spend the rest of its time on—it's worth saying that classes are really the ultimate type in Kotlin. A class provides you a way to collect data and work with that data in a specific manner, and to model things in the world: numbers, sentences, objects like cars, people, even abstract ideas like a radio wave or a decision.

Classes also give you a pointer into something that's quite important in Kotlin: type safety. *Type safety* refers to the degree to which a programming language keeps you from making mistakes related to assigning one type (like a number) to a variable that should only hold letters. There's nothing as frustrating as treating a variable like it's all letters, and things in your program break because it turns out that that variable actually contains an instance of Person (or Car, or User, or something else that doesn't at all act like letters).

And that's where classes and objects are so key to type safety. Your Person is now strongly typed; you can't create an integer and shove a string into it. More specifically, you can't create a Car and shove it into a Person. You're going to see a lot more about this as the chapter goes on. For now, though, just think of objects as a really powerful way to be sure a variable (whether by val or by var) has exactly in it what you need.

> **NOTE** *If you're wondering how Kotlin actually knows what type is allowed for a variable, keep reading. Much more on that shortly.*

KOTLIN HAS A LARGE NUMBER OF TYPES

Like any programming language, Kotlin supports lots of basic data types. You can define integers (Int), sequences of letters (String), decimal numbers (Float), and a lot more. Let's take a blazing-fast run through the basic types and then start putting them to use.

Numbers in Kotlin

Kotlin gives you four options for integers, largely varying based on range. These are shown in Table 2.1.

TABLE 2.1: Kotlin Types for Integers (Nondecimal Numbers)

TYPE	SIZE (BITS)	MINIMUM VALUE	MAXIMUM VALUE
Byte	8	−128	127
Short	16	−32,678	32,767
Int	32	−2,147,483,648 (-2^{31})	2,147,483,647 ($2^{31} - 1$)
Long	64	−9,223,372,036,854,775,808 (-2^{63})	9,223,372,036,854,775,807 ($2^{63} - 1$)

Kotlin will largely take care of figuring out which type to use when you don't declare that type explicitly. In that case, it's going to look at the value you're assigning the variable, and make some assumptions. This is a pretty important concept called *type inference*, and it's something we're going to talk about in a lot more detail in Chapter 6.

If you create a variable and assign it a number that fits into Int, then Int will be used. If the number is outside the range of Int, the variable will be created as a Long:

```
val someInt = 20
val tooBig = 4532145678
```

So here, someInt will be an Int. tooBig is too big to fit into an Int so it will be a Long. You can also force a variable to be a Long by adding a capital L to the value:

```
val makeItLong = 42L
```

Kotlin gives you a couple of options for decimals, as well: two, again, based on precision and size. Those are listed in Table 2.2.

TABLE 2.2: Kotlin Types for Decimal Numbers

TYPE	SIZE (BITS)	SIGNIFICANT BITS	EXPONENT BITS	DECIMAL DIGITS
Float	32	24	8	6–7
Double	64	53	11	15–16

Assignment here works a bit unexpectedly. Decimal variables will be Double unless you tell Kotlin to use Float by using the f or F suffix:

```
val g = 9.8
val theyAllFloat = 9.8F
```

Letters and Things

If you want to represent a single character—including special characters like backslash or a carriage return—you use the Char type:

```
val joker = 'j'
```

A character should be enclosed in a set of single quotes. You can also enclose special characters in quotes and precede the character code with a backslash. For instance, tab is \t, a line feed is \n, and a backslash is itself \\:

```
val special = '\n'
```

> **NOTE** *Backslash is weird because it is itself an escape character. To get an actual backslash, you use the escape character (\) and then another backslash (\), which gets you \\.*

For sequences of characters, you likely want a `String`. You can create a string with text enclosed in double quotes:

```
val letters = "abcde"
```

Simple enough! But note that a single letter in double quotes is a `String`, while a single letter in single quotes is a `Char`:

```
val letter = "a"
val notStringy = 'a'
```

Truth or Fiction

For true or false values, you can use a `Boolean`. You can assign to a `Boolean` either true or false:

```
val truthful = true
val notTrue = false
```

However, you *cannot* use 0 or 1 as you can in some languages. A 0 is an `Int` (or a `Long`, or even a `Float`):

```
val notBoolean = 0
val zeroInt = 0
val zeroLong = 0L
val zeroFloat = 0F
```

A 0 is never a `Boolean`, though.

Types Aren't Interchangeable (Part 1)

Most of these types are likely unsurprising. They're typical for most languages. What will surprise you a bit is how far Kotlin will go to ensure you don't get your types wrong. Remember, Kotlin is strongly typed, and the language takes that pretty seriously.

To see this in action, let's add some basic information to the `Person` class. Listing 2.3 shows a slightly modified version of this class with some more properties and types.

LISTING 2.3: Person with extra properties

```
package org.wiley.kotlin.person

class Person(var firstName: String, var lastName: String) {
    var fullName: String
    var height: Float
    var age: Int
    var hasPartner: Boolean

    // Set the full name when creating an instance
    init {
        fullName = "$firstName $lastName"
    }
```

LISTING 2.3 *(continued)*

```
    override fun toString(): String {
        return fullName
    }
}
```

You'll notice that each of these new properties has a type: `String` or `Float` or `Int`. Simple enough.

Now, there are some interesting things about the types of these variables, but before getting to that, there's actually a new problem in `Person` (in addition to that last name thing that still needs to be dealt with).

You Must Initialize Your Properties

Kotlin requires that you initialize your properties. If you try to compile the `Person` class in Listing 2.2, you're going to get an error like this:

```
Error:(6, 5) Kotlin: Property must be initialized or be abstract
```

Let's leave the abstract piece of that for later. The easiest fix is to update the `Person` constructor to require all of this information. That's pretty straightforward:

```
class Person(var firstName: String, var lastName: String,
             var height: Float, var age: Int, var hasPartner: Boolean) {
    var fullName: String

    // Set the full name when creating an instance
    init {
        fullName = "$firstName $lastName"
    }

    // other methods follow
}
```

Now you have to pass in these properties, so they'll pass compilation. Also note that by putting them in the constructor, you don't have to declare them within the class body.

> **NOTE** *You can also create additional versions of a* `Person` *constructor if you want to only take in parts of this information on instance creation. We'll come back to that topic in more detail in Chapter 4.*

Alternatively, you could assign these properties values in the `init` method, as you did with `fullName`. However, there's no real way to come up with values for height, age, or whether the person has a partner without taking in input, so this is the more sensible approach.

Now you're ready to go back to `main` and do some type-related work.

Types Aren't Interchangeable (Part 2)

First, take a look at PersonApp, and specifically your main function. Most IDEs will see that you've updated the constructor and give you some visual cues as to what properties you're currently passing into Person. Note in Figure 2.4 that IntelliJ has identified the firstName and lastName properties being passed in so far.

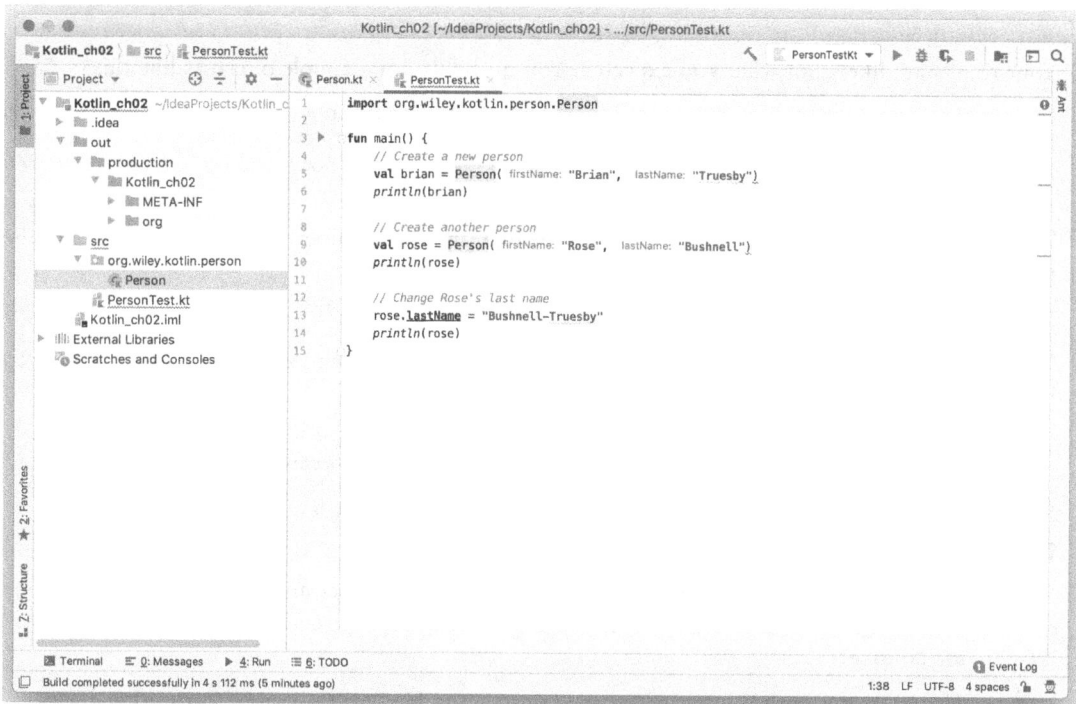

FIGURE 2.4: IDEs will help you keep constructor properties straight.

Go ahead and make the changes to your function to pass in some values to fill out the new constructor for Person. You'll need to add a height (I chose to use inches, but any unit of measurement works fine), an age, and whether the person has a partner. Listing 2.4 shows the result.

LISTING 2.4: Filling out your Person instances

```
import org.wiley.kotlin.person.Person

fun main() {
    // Create a new person
    val brian = Person("Brian", "Truesby", 68.2, 33, true)
    println(brian)

    // Create another person
    val rose = Person("Rose", "Bushnell", 56.8, 32, true)
```

LISTING 2.4 *(continued)*

```
    println(rose)

    // Change Rose's last name
    rose.lastName = "Bushnell-Truesby"
    println(rose)
}
```

Your results are likely not what you expected. You probably got a couple of errors like this:

```
Error:(5, 44) Kotlin: The floating-point literal does not conform to the expected
type Float
Error:(9, 43) Kotlin: The floating-point literal does not conform to the expected
type Float
```

What in the world does this mean? If you find the line (the first number in the parentheses) and the place on that line (position 44), you'll see that the error occurs here:

```
val brian = Person("Brian", "Truesby", 68.2, 33, true)
```

You're passing in a decimal for height—in this case 68.2—which seems right. Here's the constructor in `Person`:

```
class Person(var firstName: String, var lastName: String,
             var height: Float, var age: Int, var hasPartner: Boolean) {
```

So what gives? The problem is that Kotlin is not converting types. A few things are actually happening here:

1. You give Kotlin the value 68.2. Kotlin sees a decimal number and automatically uses the `Double` type. This is important! *Anytime* a decimal is not given an explicit type, Kotlin will use `Double`.

2. The `Double` is passed into the `Person` constructor. However, `Person` is expecting a `Float`.

3. Kotlin will *not* try to convert between these two types—even though, in this case, they are compatible! Instead, Kotlin insists on strong typing, and throws an error.

You Can Explicitly Tell Kotlin What Type to Use

You have two ways to fix this. The easiest is to tell Kotlin that you want the value you're passing in to be treated as a `Float`; you can do that by putting a capital `F` after the number:

```
val brian = Person("Brian", "Truesby", 68.2F, 33, true)
println(brian)

// Create another person
val rose = Person("Rose", "Bushnell", 56.8F, 32, true)
println(rose)
```

In this case, you're sending `Person` what it wants: a `Float`. That passes the type checking and things are OK again.

Try to Anticipate How Types Will Be Used

While that solution works, it's a bit inelegant. It is going to require every user of your `Person` class—you included—to add that F to decimals or to use the `Float` type for variables passed into a constructor. And, as you just learned, Kotlin uses `Double` as the default decimal type.

A better solution is to realize that if Kotlin defaults to `Double`, maybe you should, too. (At least, unless you have a really good reason not to.) So change your `Person` constructor to take in what most users will pass—a `Double` created from a typed-in number:

```
class Person(var firstName: String, var lastName: String,
             var height: Double, var age: Int, var hasPartner: Boolean) {
```

Easy enough! And, because you've anticipated how users will create a new instance of `Person`, you'll annoy your friendly developer comrades a lot less.

> **NOTE** *Don't forget to remove the* F *from the two numbers in your* main *function. If you don't, you'll get another error—this time from trying to pass a* Float *into a constructor that expects a* Double.

It's Easy to Break Kotlin (Sort of)

The title of this chapter is "It's Hard to Break Kotlin," which may seem like the opposite of what you're seeing. Without much effort, you've managed to get Kotlin to throw a lot of different errors, and Kotlin seems quite resistant to helping you out with converting between types.

That said, Kotlin is actually breaking *before* your code is running in some production system, serving an API for a mobile app, or running on a phone or just handling web requests. In other words, Kotlin is making you do some extra work at compile time—when you're writing code—to avoid potential problems when the program is running and *needs* to work.

Put another way, the chapter might be better titled, "It's Hard to Break Kotlin When It's Running in Production Because It's Going to Make You Be Really Specific about Types When You're Writing that Code Before It Gets to Production." Of course, that's really not a title that anyone is going to let get through editing and proofing, so let's go with "It's Hard to Break Kotlin" and trust that you know what that really means.

OVERRIDING PROPERTY ACCESSORS AND MUTATORS

With what you know about types, and strong types in particular, you're finally ready to get back to that annoying little detail of `Person`: if you set the last name outside the `Person` constructor, printing the class instance gives the original last name, not the modified one. As a refresher, here's the code in question:

```
// Create another person
val rose = Person("Rose", "Bushnell", 56.8, 32, true)
println(rose)
```

```
// Change Rose's last name
rose.lastName = "Bushnell-Truesby"
println(rose)
```

The result? This:

```
Rose Bushnell
Rose Bushnell
```

What really needs to happen here? Well, every time that a last name—or first name, for that matter—is updated, the full name of the person that the instance represents needs to be updated. Seems simple enough, right?

Well, hold on, because this is going to get fairly complicated really fast. You're about to enter the somewhat unusual world of overriding mutators and accessors in Kotlin.

Custom-Set Properties Can't Be in a Primary Constructor

What a mouthful! It's true, though: because we want to take control of what happens when a first or last name is changed, we're going to have to make a lot of changes to how Person is constructed. Basically, we have to override the accessor for firstName and lastName. To do that, we cannot have those properties initialized in the Person primary constructor.

> **NOTE** *You may have noticed the term "primary constructor" here. Think of that simply as "the constructor" for now. Later, you'll see that Kotlin lets you define multiple constructors, and the primary one is the one defined on the first line of the class—the one you already have.*
>
> *More on this later, though, so don't worry too much about primary and secondary constructors for now.*

Move Properties Out of Your Primary Constructors

You've actually just stumbled onto a fairly standard best practice in Kotlin: it's often a good idea to move property definitions out of the primary constructor. You'll recall that you started like this way back in Chapter 1:

```
class Person(firstName: String, lastName: String) {
```

Then, before you added other properties, you added the var keyword to each listed data input:

```
class Person(var firstName: String, var lastName: String) {
```

This made firstName and lastName properties, with automatically generated accessors and mutators. The problem is that, while expedient at the time, this has now come back to bite you.

To undo this, simply remove `var` from all of the properties in the current version of the `Person` constructor:

```
class Person(firstName: String, lastName: String,
             height: Double, age: Int, hasPartner: Boolean) {
```

Now try to run your `PersonApp`, and you're going to see an error discussed earlier:

```
Error:(13, 10) Kotlin: Unresolved reference: lastName
```

You should recall that because `lastName` is not a property anymore (because there's no var or val keyword in the constructor definition), Kotlin does *not* create accessors or mutators. That means that this line in `main` now won't work:

```
rose.lastName = "Bushnell-Truesby"
```

So there's a bit of an apparent conflict that has arisen:

➤ You can't customize accessors or mutators if a property is declared in a class's constructor.

➤ If a property isn't declared in a class's constructor, and it's not declared elsewhere in the class, you don't get an automatic accessor or mutator.

Initialize Properties Immediately

Now, just to make this a bit more complicated, there's another rule you've run across, too:

➤ Any property not defined in a constructor must be initialized on instance creation.

To remember how this shows up, go ahead and define all the properties declared in your constructor, but this time do it just *under* the constructor line, similar to how you defined `lastName`:

```
class Person(firstName: String, lastName: String,
             height: Double, age: Int, hasPartner: Boolean) {
    var fullName: String
    var firstName: String
    var lastName: String
    var height: Double
    var age: Int
    var hasPartner: Boolean

    // Set the full name when creating an instance
    init {
        fullName = "$firstName $lastName"
    }
```

Compile, run, and you'll get another error you've seen before, but this time with all of your properties:

```
Error:(6, 5) Kotlin: Property must be initialized or be abstract
Error:(7, 5) Kotlin: Property must be initialized or be abstract
Error:(8, 5) Kotlin: Property must be initialized or be abstract
Error:(9, 5) Kotlin: Property must be initialized or be abstract
Error:(10, 5) Kotlin: Property must be initialized or be abstract
```

This cropped up with `firstName`, and we fixed it by setting that value in your code's init block. We need to do something somewhat similar here: we need to assign to each new property the value that is passed into the constructor. So the property `firstName` should be given the value passed into `firstName` that's declared as an input to the `Person` constructor.

> **NOTE** *If that sentence seemed confusing, you're in good company. Two* `first-Name` *pieces of information are floating around here: the one passed into the constructor, and the property name. That's a problem we'll fix shortly.*

Kotlin gives you a way to this assignment but it's going to look a little odd. Basically, the values that come into your constructor are available to your properties at creation time, and you can assign them directly to those properties. That's a little easier said than described, so take a look at the new version of `Person` shown in Listing 2.5.

LISTING 2.5: Assigning constructor information to properties

```
package org.wiley.kotlin.person

class Person(firstName: String, lastName: String,
             height: Double, age: Int, hasPartner: Boolean) {
    var fullName: String
    var firstName: String = firstName
    var lastName: String = lastName
    var height: Double = height
    var age: Int = age
    var hasPartner: Boolean = hasPartner

    // Set the full name when creating an instance
    init {
        fullName = "$firstName $lastName"
    }

    override fun toString(): String {
        return fullName
    }
}
```

Take just the line dealing with `firstName`:

```
var firstName: String = firstName
```

A property is declared, and it's both read and write, because `var` is used. (Remember, if you wanted this to be a read-only property, you'd use `val`.) That property is given a name—`firstName`—and then a type, `String`. Then, the new property is assigned a value. In this case, that value is whatever is passed into the constructor through `firstName`.

You can now compile your code and run it, and it works again! (Although there's still that last name issue to fix. But you're getting closer.)

Try to Avoid Overusing Names

This code is still pretty confusing, though. You have properties named the same as inputs to your constructor, which is legal, but a pain to understand and follow. One easy solution, and a common one in the Kotlin community, is to use an underscore for non–property value names passed into constructors. Listing 2.6 shows this change; it's purely cosmetic but really cleans up the readability of the class.

LISTING 2.6: Clarifying property names and constructor inputs

```kotlin
package org.wiley.kotlin.person

class Person(_firstName: String, _lastName: String,
             _height: Double, _age: Int, _hasPartner: Boolean) {
    var fullName: String
    var firstName: String = _firstName
    var lastName: String = _lastName
    var height: Double = _height
    var age: Int = _age
    var hasPartner: Boolean = _hasPartner

    // Set the full name when creating an instance
    init {
        fullName = "$firstName $lastName"
    }

    override fun toString(): String {
        return fullName
    }
}
```

> **NOTE** *Because Kotlin is so strongly typed and has such unique functionality— multiple constructors, properties versus method inputs, and more—using these coding best practices will really help separate your code from the code of those who are less experienced and savvy.*

Override Mutators for Certain Properties

You're now finally ready to do what we set out to do: override how changing a first name or last name works. First, you need to see how a typical mutator is defined.

Here's some more rather Kotlin-specific code that defines a custom mutator; in this case, it's for `firstName`:

```kotlin
var firstName: String = _firstName
    set(value) {
        field = value
    }
```

The `set` keyword tells Kotlin that this is a mutator. Additionally, Kotlin assumes that the mutator is for whatever property was just defined.

> **WARNING** *Kotlin uses a property's definition line in a file to determine that, if there's a custom mutator (or accessor), that definition is on the very next line. That's a big deal, and worth noting.*

Then, `value` represents the value coming into the mutator. So in the following example, `value` would come to the `firstName` mutator as "Bobby":

```
// Change Brian's first name
brian.firstName = "Bobby"
```

Now here's where things look a little odd. There's this line:

```
field = value
```

What's going on there? What the heck is `field`? Well, it's the property being mutated. It's what's called a *backing field*. So here, `field` references the backing field for the property, which is `first-Name` (because that's the property just defined). That backing field is assigned the value passed into the mutator—in our example, that's the `String` "Bobby."

The final piece here is understanding why you can't say something like this:

```
var firstName: String = _firstName
    set(value) {
        firstName = value
    }
```

This will actually compile, but it will give you a horrible error when you try to run and use this class. The problem is that when Kotlin sees this:

```
firstName = value
```

it interprets that as "run the code that mutates `firstName`." But that code is the code *already running*. So it calls itself—a concept called *recursion*—and that same line runs again. And again. And again. And. . . well, you get the idea.

By using the backing field `field`, you avoid this recursion, and things behave. Go ahead and create a custom mutator for `firstName` and `lastName`; things should look like Listing 2.7 when you're finished.

LISTING 2.7: Defining custom mutators for lastName and firstName

```
package org.wiley.kotlin.person

class Person(_firstName: String, _lastName: String,
            _height: Double, _age: Int, _hasPartner: Boolean) {
    var fullName: String
    var firstName: String = _firstName
```

```
        set(value) {
            field = value
        }
    var lastName: String = _lastName
        set(value) {
            field = value
        }
    var height: Double = _height
    var age: Int = _age
    var hasPartner: Boolean = _hasPartner

    // Set the full name when creating an instance
    init {
        fullName = "$firstName $lastName"
    }

    override fun toString(): String {
        return fullName
    }
}
```

> **WARNING** *The good news about Listing 2.7 is that you can literally copy the mutator code from* firstName *to use again for* lastName. *Since* field *applies to whatever backing field is being used, it works for both properties. The bad news is that skimming Kotlin code can sometimes be hairy, because you'll see lots of blocks that look similar—just like these.*

CLASSES CAN HAVE CUSTOM BEHAVIOR

What we need is a way to define some custom behavior. Specifically, when a first name or last name of a person is changed, their full name should also be changed. With custom mutators, part of this work is done; we have a *place* to call custom code, but no custom code to call.

Define a Custom Method on Your Class

What we need, then, is a new class method. Let's call it updateName(), and have it update the full-Name property. Then, we can call updateName() every time a name needs to be changed. There's really nothing magical here, as you already have created a method with fun. Here's what you really want:

```
fun updateName() {
    fullName = "$firstName $lastName"
}
```

This is actually the exact same code that already exists in your Person's init block. But that code isn't needed anymore! Instead, you can call updateName() within your init block. Listing 2.8 shows you what Person should look like when you're finished.

> **LISTING 2.8:** Creating a new property method to update the fullName variable

```
package org.wiley.kotlin.person

class Person(_firstName: String, _lastName: String,
             _height: Double, _age: Int, _hasPartner: Boolean) {
    var fullName: String
    var firstName: String = _firstName
        set(value) {
            field = value
        }
    var lastName: String = _lastName
        set(value) {
            field = value
        }
    var height: Double = _height
    var age: Int = _age
    var hasPartner: Boolean = _hasPartner

    // Set the full name when creating an instance
    init {
        updateName()
    }

    fun updateName() {
        fullName = "$firstName $lastName"
    }

    override fun toString(): String {
        return fullName
    }
}
```

Every Property Must Be Initialized

This doesn't look a lot different. However, there's a problem. Compile this code and you're going to get an error that, by now, is probably becoming a bit familiar:

```
Error:(5, 5) Kotlin: Property must be initialized or be abstract
```

What's going on here? Well, it's a little bit of a pain. Remember, any property, *must* either be assigned an initial value when they are declared (like `firstName` and `lastName`, for example), or be assigned a value in the `init` block.

Now, it *looks* like that's what is happening, but Kotlin doesn't quite follow your code the same way that you do. While it is possible to see that when `init` calls `updateName()`, then `fullName` will get a value, Kotlin just sees that there's no assignment to `fullName` in `init` and throws an error.

Assign an Uninitialized Property a Dummy Value

The easiest fix here is actually quite ... easy. You can simply get around Kotlin's insistence on property initialization by assigning it an empty value at creation, such as an empty `String`: `""`. Just add this to your class:

```
class Person(_firstName: String, _lastName: String,
             _height: Double, _age: Int, _hasPartner: Boolean) {
    var fullName: String = ''
```

Kotlin will now stop complaining! You've initialized `fullName`, so there's not a problem. However, this is a bit hacky. It defeats the purpose of Kotlin's checking (something we'll come back to in a moment). It also builds a hidden dependency in your code. Look at the `init` method again:

```
    // Set the full name when creating an instance
init {
    updateName()
}
```

Even with the comment, there's nothing except your own memory and understanding of how `Person` works—and this chapter—that tells you that you *have* to call `updateName()` in `init`. If you don't, then `fullName` will not get initialized, and that, in turn, will mess up the `toString()` function. Bad news!

If this seems unlikely, it's not. When you're writing code, you know what that code is supposed to do. But leave that code for a few weeks, or months, and you often forget how it worked, let alone why you wrote it that way! Worse, but just as likely, *someone else* comes back to it later and has no idea how it works.

In both cases, it would be quite plausible that someone well-meaning forgets that `updateName()` must be called in `init`, removes or moves it, and problems ensue. This is really a fragile solution.

Tell Kotlin You'll Initialize a Property Later

Another easy solution is to explicitly tell Kotlin that you are going to initialize a property later. It's sort of the equivalent of saying, "Look, Kotlin, trust me. I promise I'll take care of this." To do this, you preface a property declaration with `lateinit`, a keyword that means what it sounds like—you'll initialize the property in question later:

```
class Person(_firstName: String, _lastName: String,
             _height: Double, _age: Int, _hasPartner: Boolean) {
    lateinit var fullName: String
```

Remove the assignment to `""` and add `lateinit`. Now, your class will compile again! No problem.

> **WARNING** `lateinit` *has some pretty specific limitations. It only works with* `var`, *not* `val`, *and the properties can't be declared within the constructor itself. You also can't create a custom accessor or mutator for a* `lateinit`*-declared property. So it's not a fix-all. In fact, it's not even a fix in this case, really.*

While it might seem this solves your problem, that's not actually true. You still have the same problem as assigning an empty string to `fullName`: you've built a silent dependence on `updateName()` being called in `init`. Kotlin is no longer checking for you, because you said, "I'll take care of it."

So this is still a fragile solution, and in some cases, even more dangerous than just using a dummy value to start with.

Assign Your Property the Return Value from a Function

It's worth taking a minute to remember what the actual problem here is. You need to ensure that Kotlin sees an assignment of a value to `fullName` in one of three legal places:

➤   In the class's primary constructor

➤   In the property's declaration

➤   In the class's init block

You've also seen that simply calling another function—like `updateName()`—and letting that function do the assignment won't cut it.

The first thing we need to do is change our function. Why? Because it currently doesn't return a value—and that means that you can't use the function as the right side of an assignment. In other words, you can't get the all-important:

```
var fullName: String = // something here... like a function!
```

So change `updateName()` to actually return a value. While you're at it, let's also change the name to better reflect what it does now:

```
fun combinedNames(): String {
    return "$firstName $lastName"
}
```

Now, you need a little cleanup:

1.   Remove the reference to `updateName()` from `init`.

2.   Assign `fullName` the return value from this function.

3.   While you're at it, remove the `init` block entirely.

4.   Move the declaration of the `fullName` property from the first line in `Person` to after all the other properties are declared.

> **WARNING** *Don't miss that last item! Now that you're using the values of* `firstName` *and* `lastName` *to initialize* `fullName`, *the order of property declaration matters. You need* `firstName` *and* `lastName` *to get assigned the values from the primary constructor before* `fullName` *is declared and* `combinedNames()` *is run.*

You can see this all together in Listing 2.9.

Listing 2.9 Getting the fullName property correct (partially)

```kotlin
package org.wiley.kotlin.person

class Person(_firstName: String, _lastName: String,
             _height: Double, _age: Int, _hasPartner: Boolean) {

    var firstName: String = _firstName
        set(value) {
            field = value
        }
    var lastName: String = _lastName
        set(value) {
            field = value
        }
    var height: Double = _height
    var age: Int = _age
    var hasPartner: Boolean = _hasPartner
    var fullName: String = combinedNames()

    fun combinedNames(): String {
        return "$firstName $lastName"
    }

    override fun toString(): String {
        return fullName
    }
}
```

Sometimes You Don't Need a Property!

Unbelievably, after all this work, the output from your code is *still incorrect* if a last name is changed after object instance creation. One simple fix here is to update the two custom mutators, for first-Name and lastName, to update fullName, like this:

```kotlin
    var firstName: String = _firstName
        set(value) {
            field = value
            // Update the full name
            fullName = combinedNames()
        }
    var lastName: String = _lastName
        set(value) {
            field = value
            // Update the full name
            fullName = combinedNames()
        }
```

Now, finally, things will work! The rather unexciting output after all this work should look like this:

```
Brian Truesby
Rose Bushnell
Rose Bushnell-Truesby
```

But before you pop the champagne, take a minute to think this through. Is this really a good solution? You have several problems here, all of which are nontrivial. See if these make sense; by now, they all should:

➤ You've got the same piece of code—the assignment to `fullName`—in both the `firstName` and `lastName` mutator. In general, anytime you see the same code twice, that's not great. Try to avoid repeating code, as it becomes error-prone.

➤ You're going to a *lot* of work to set the value of a property that has *no special behavior*. It's just the combination of the first and last names of the instance!

➤ You *already* have a function that gives you the full name; it's called `combinedNames()`.

A much better approach, believe it or not, is to just remove `fullName` altogether. In fact, there's a *lot* that can be removed here.

Let's use `combinedNames()` to handle the combination of `firstName` and `lastName`. Then you can remove the declaration of the `fullName` property altogether. But if you do that, then you can actually remove the custom mutator for `firstName` and `lastName`, too!

And go one further; rename `combinedNames()` to `fullName()`, and then update `toString()` to just use that. After all that, you should have something like Listing 2.10, which is pretty minimal!

LISTING 2.10: A much simpler version of Person

```kotlin
package org.wiley.kotlin.person

class Person(_firstName: String, _lastName: String,
             _height: Double, _age: Int, _hasPartner: Boolean) {

    var firstName: String = _firstName
    var lastName: String = _lastName
    var height: Double = _height
    var age: Int = _age
    var hasPartner: Boolean = _hasPartner

    fun fullName(): String {
        return "$firstName $lastName"
    }

    override fun toString(): String {
        return fullName()
    }
}
```

Compile and then run `PersonApp` and you should get correct results—both before and after updating the last name for the `rose` instance.

TYPE SAFETY CHANGES EVERYTHING

The big takeaway here is that when a language seeks to really enforce type safety—as Kotlin does—then nearly everything in the language gets affected. Properties are required to have values, and have them early, so that Kotlin can ensure the correct types are used. The mutability of variables is tightly controlled—in Kotlin's case, through `var` and `val` being different—so that Kotlin can determine when it needs to keep up with a variable's type over time, and the values assigned to that type. And Kotlin doesn't convert types for you. All of these nuances to Kotlin are keeping type safety intact and making it much harder to break Kotlin at runtime.

Believe it or not, there is a lot more to Kotlin's type safety: things you'll learn in future chapters. Kotlin has a lot of specific rules around `null` values, checking for `null`, and the idea of *null safety*: keeping your code free from null pointer exceptions.

> **NOTE** *If you're not familiar with null pointers, or have never seen a* `NullPointerException` *in, for example, Java, then count yourself lucky. If you have, though, they're often the source of very difficult bugs. Kotlin attempts to add some extra work at the coding layer in favor of getting rid of those annoying and hard-to-find bugs. It's a trade-off, but one that works quite well for Kotlin.*

WRITING CODE IS RARELY LINEAR

If you take a step back and consider what you've done over these last two chapters, it's both cool and potentially a little confusing. The cool part is even if you've never written a line of Kotlin before, you've now got a functioning class that exercises some of Kotlin's key and relatively unique features: objects, construction, and overriding superclasses (more on that later), as well as the somewhat unusual requirements Kotlin enforces around property value assignments.

What could be a bit confusing, though, is how you got to the terse code in Listing 2.10. You added properties, removed them, added a couple of custom mutators, removed those, and ultimately deleted nearly as much code as you added. Wouldn't it have been easier to just jump right to Listing 2.10?

Well, it might have, but it wouldn't have been that helpful in getting you to see how Kotlin works. Arguably just as important, it wouldn't be realistic. Seasoned developers are constantly going through this same process: trying things out, then adding something new, having to go back and change old code to accommodate new code, tweaking, and removing as much as is added. It's common to refactor—to restructure code to make it less redundant, more efficient, more organized, and generally just better.

That's really the point if you're learning Kotlin, too: to get more than just a syntactical understanding. You want to be a good developer? Keep adding things, removing things, and taking a fresh look to make sure your code is the *right* code.

3

Kotlin Is Extremely Classy

WHAT'S IN THIS CHAPTER?

➤ Every class has an `equals()` and `hashCode()` method

➤ Different types of object and object instance comparison

➤ Overriding `equals()` and `hashCode()` in meaningful ways

➤ The importance of `equals()` and `hashCode()` in comparison operations

➤ Why class basics matter

OBJECTS, CLASSES, AND KOTLIN

It is not an overstatement to say that Kotlin is entirely about objects, which necessarily means it is all about classes. Since an object is just a specific instance—an *instantiation*—of a class, any language that makes heavy usage of objects is going to have a lot of class-based logic.

Put more simply, if you want to be a good Kotlin programmer, you need to understand classes, as you'll be using them a lot. In fact, most Kotlin applications have *at most* a single nonclass involved: the main function that you've already seen.

In this chapter, you're going to get more into the weeds with classes, and really understand what they come with "out of the box" in Kotlin, and how to use them more effectively. Literally everything else you do will be based on this, so take your time, and try not to skip the boring parts.

> **NOTE** *It may seem an odd thing for a writer to say "the boring parts," but some boring parts are definitely coming. Not everything about classes in Kotlin is particularly exciting or even fun. That said, you've got to get a hold of these things to know how to use classes to maximum advantage. So. . . boring? Yup. Essential? Also yup.*

ALL CLASSES NEED AN EQUALS(X) METHOD

The first thing you should realize is something you've already used: each class in Kotlin has some methods already defined. You saw this in the `Person` class when you overrode `toString`.

> **NOTE** *There is actually a lot more going on with* `toString` *and similar methods than it just "existing" for every class. You'll get much more into these methods— which are actually inherited from another class—in Chapter 5. For now, it's safe to think of these as predefined methods.*

Classes also come with two more methods "for free" like this: `equals()` and `hashCode()`. You'll find that you use `equals()` quite a bit—like `toString()`—and `hashCode()` very rarely in typical Kotlin programs.

Equals(x) Is Used to Compare Two Objects

The `equals()` method is actually usually written like `equals(x)`, which indicates that it takes in a parameter (represented by x). The parameter is another object, and the method returns either true or false. True indicates that the object itself (the one that you called `equals(x)` on) and the object passed into the method are identical. Listing 3.1 shows how you could use this method with `Person` instances.

LISTING 3.1: Using equals(x) to compare two instances

```kotlin
import org.wiley.kotlin.person.Person

fun main() {
    // Create a new person
    val brian = Person("Brian", "Truesby", 68.2, 33, true)
    println(brian)

    // Create another person
    val rose = Person("Rose", "Bushnell", 56.8, 32, true)
    println(rose)

    // Change Rose's last name
    rose.lastName = "Bushnell-Truesby"
    println(rose)

    if (rose.equals(brian))
        println("It's the same person!")
    else
        println("Nope... not the same person")
}
```

As you'll see in a minute, though, the default version of this method almost never does what you want it to.

> **WARNING** *Take a moment to realize that* equals(x) *compares two instances of a class, rather than two actual classes. And in general, it will always return false if the two objects aren't instances of the same exact class.*
>
> *This of course makes sense: how could an instance of* Person *ever be considered equal to an instance of* Automobile?

If you run Listing 3.1, you'll get just what you'd expect:

```
Brian Truesby
Rose Bushnell
Rose Bushnell-Truesby
Nope... not the same person
```

But there's a problem. To see it in action, make this addition:

```
if (rose.equals(brian))
    println("It's the same person!")
else
    println("Nope... not the same person")

val second_brian = Person("Brian", "Truesby", 68.2, 33, true)
if (second_brian.equals(brian))
    println("It's the same Brian!")
else
    println("Nope... not the same Brian")
```

Now run your code, and the output may not be what you'd expect:

```
Brian Truesby
Rose Bushnell
Rose Bushnell-Truesby
Nope... not the same person
Nope... not the same Brian
```

So what's going on? Why are two object instances with the exact same property values not identical? It has to do with the way that Kotlin's default implementation of equals(x) works: the equality is based on memory location.

The equality evaluated by equals(x) by default is the equality of the actual object instance in memory compared to another object instance's location in memory. In other words, equals(x) does not care if two objects have the exact same value; it only cares if the two instances are the actual same bit of memory. To see this in action, add some more code:

```
val copy_of_brian = brian
if (copy_of_brian.equals(brian))
    println("Now it's the same Brian!")
else
    println("Nope... still not the same Brian")
```

This time, you'll get a match:

```
Nope... not the same person
It's the same Brian!
Now it's the same Brian!
```

Now, `brian` is considered equal to `copy_of_brian`, because they are actually the same object instance! You have two variables, but they both point to the same pieces of actual memory. That is different than `second_brian`, which is another instance of `Person`—and therefore a different piece of memory—that just happens to have the same values as `brian` (and by extension, `copy_of_brian`).

Override equals(x) to Make It Meaningful

So, two objects that are identical in terms of property values, but aren't actually a single object in memory, will be considered unequal. That is something you can fix, though, and you should!

Your first step is to decide what equality means for instances of the `Person` object. The easiest solution here is to use the combination of first name and last name.

> **WARNING** *Using first name and last name is actually not a great idea, since you can already probably recall two people you know with the same first and last name. A better idea would be something like Social Security number. For the purpose of this example, though, let's consider first name and last name combined as unique.*

With this in mind, it's now easy to build a useful overridden version of `equals(x)`. Here's a good start:

```
override fun equals(other: Any?): Boolean {
    return (firstName.equals(other.firstName)) &&
        (lastName.equals(other.lastName))

}
```

This has a couple of new pieces of syntax to get your head around, and then there's an actual error here. You'll need to understand both to get this completely working.

First, the syntax. You've already seen how to override a function, so the `override` keyword is nothing new. As for the declaration of `equals(x)`, for now that's just how Kotlin defines that method, so it's always going to be the same:

```
override fun equals(other: Any?): Boolean {
```

> **NOTE** *Some really interesting things are related to that* `Any?` *bit, but hold tight. That will make more sense in just a minute.*

This function returns a `Boolean`, and that's what the long line of code does:

```
return (firstName.equals(other.firstName)) &&
        (lastName.equals(other.lastName))
```

First, it compares the `firstName` property of the current object instance with the `firstName` property of the instance passed in. Then, it does the same comparison on `lastName` to see if this instance is equal to the `lastName` of the other instance passed in.

Then, it uses `&&` to combine those two values. `&&` takes two Boolean values—one to the left of the `&&` and one to the right—and evaluates the *logical and* of the two values. Here's how that works:

1. If the first value (the one to the left of `&&`) is false, then the entire expression is false.

2. If the first value is true, then the second value is evaluated. If it's false, the expression is false.

3. If both the first and second value are true, then the expression is true.

Here's another, longer way to write the same expression:

```
if (firstName.equals(other.firstName)) {
    if (lastName.equals(other.lastName)) {
        return true
    } else {
        return false
    }
} else {
    return false
}
```

Of course, programmers like things brief, so `&&` is usually the way to go.

> **WARNING** *In addition to this being a lot longer, it's also not nearly as clear and simple to read.*

The problem is this code won't compile. If you try, you'll get errors that look like this:

```
Error:(26, 44) Kotlin: Unresolved reference: firstName
Error:(26, 82) Kotlin: Unresolved reference: lastName
```

What's going on here? Well, it has to do with that definition of `equals(x)`:

```
override fun equals(other: Any?): Boolean {
```

Specifically, the issue is that `equals(x)` takes in an object instance called `other` of type `Any?`. `Any` in Kotlin is the base class, and what you can take this to mean for now is that literally anything can be passed into this method, and Kotlin will happily pass that into your `equals(x)` method.

In the case of `Person`, then, Kotlin is telling you through those errors that, "It's not for sure that you're getting a `Person` instance, so there might not be properties called `firstName` and `lastName`." That's where the unresolved reference is coming into play: Kotlin can't be sure those references exist.

Think about that for a second. That means, then, that this code that sees if two Person instances are equal is valid:

```
val brian = Person("Brian", "Truesby", 68.2, 33, true)
val rose = Person("Rose", "Bushnell", 56.8, 32, true)

if (rose.equals(brian))
    println("It's the same person!")
else
    println("Nope... not the same person")
```

But it also means that this code is *also* valid:

```
val brian = Person("Brian", "Truesby", 68.2, 33, true)
val ford = Car("Ford", "Ranger")

if (ford.equals(brian))
    println("It's the same person!")
else
    println("Nope... not the same person")
```

This should seem weird, right? But it's completely legal: there's nothing that prevents you from comparing two totally different object type instances to each other.

Every Object Is a Particular Type

What you need, then, is a way to ensure that you're dealing with an instance of Person. If you're not, then you can safely return false. If the object instance passed into equals(x) *is* a Person instance, then your current comparison will work.

Kotlin provides a helpful keyword for this: is. You can see if a particular instance is a certain type of object by using is like this:

```
brian is Person // true
brian is Automobile // false
```

Take this into equals(x) now:

```
override fun equals(other: Any?): Boolean {
    if (other is Person) {
        return (firstName.equals(other.firstName)) &&
                (lastName.equals(other.lastName))
    } else {
        return false
    }
}
```

Kotlin will happily compile this code, because *before* you attempt to reference firstName or last-Name, your code ensures that the object is indeed a Person—which has those properties.

You can finally see all this work in action in Listing 3.2.

LISTING 3.2: Person with a working equals(x) override

```kotlin
package org.wiley.kotlin.person

class Person(_firstName: String, _lastName: String,
             _height: Double, _age: Int, _hasPartner: Boolean) {

    var firstName: String = _firstName
    var lastName: String = _lastName
    var height: Double = _height
    var age: Int = _age
    var hasPartner: Boolean = _hasPartner

    fun fullName(): String {
        return "$firstName $lastName"
    }

    override fun toString(): String {
        return fullName()
    }

    override fun equals(other: Any?): Boolean {
        if (other is Person) {
            return (firstName.equals(other.firstName)) &&
                    (lastName.equals(other.lastName))
        } else {
            return false
        }
    }
}
```

Listing 3.3 shows a version of `PersonTest` that pulls together the various snippets to test out this implementation.

LISTING 3.3: Testing out equals(x) in a few scenarios

```kotlin
import org.wiley.kotlin.person.Person

fun main() {
    // Create a new person
    val brian = Person("Brian", "Truesby", 68.2, 33, true)
    println(brian)

    // Create another person
    val rose = Person("Rose", "Bushnell", 56.8, 32, true)
    println(rose)

    // Change Rose's last name
    rose.lastName = "Bushnell-Truesby"
    println(rose)
```

LISTING 3.3 *(continued)*

```
    if (rose.equals(brian))
        println("It's the same person!")
    else
        println("Nope... not the same person")

    val second_brian = Person("Brian", "Truesby", 68.2, 33, true)
    if (second_brian.equals(brian))
        println("It's the same Brian!")
    else
        println("Nope... not the same Brian")

    val copy_of_brian = brian
    if (copy_of_brian.equals(brian))
        println("Now it's the same Brian!")
    else
        println("Nope... still not the same Brian")
}
```

The output here is just what you'd hope:

```
Brian Truesby
Rose Bushnell
Rose Bushnell-Truesby
Nope... not the same person
It's the same Brian!
Now it's the same Brian!
```

Congratulations! All that work, but you now have useful comparisons.

A Brief Introduction to Null

One quick note that you might have caught yourself: equals(x) takes in an Any object, but there's an odd ? after the Any class name:

```
override fun equals(other: Any?): Boolean {
```

That ? means that the method will accept null values. You've not dealt with null before, and it doesn't need much attention now. But this allows null to be passed into equals(x), which is important because there are times when a variable may not have a value attached, and still needs to be compared to another object instance (like your Person instances).

Fortunately, the is keyword you're using also is just fine comparing an instance to null; it will always return false, which is perfect for your overridden equals(x).

EVERY OBJECT INSTANCE NEEDS A UNIQUE HASHCODE()

Great! You've got equality covered! But there's something you're (now) missing: an overridden implementation of `hashCode()` to match. To understand why, you need to take a deeper look at `Any`.

All Classes Inherit from Any

Every Kotlin object class has `Any` as its base class. That doesn't mean much now if you're new to inheritance, but it will later. For now, it just means that the `Any` class defines behavior that *all* Kotlin classes—including the ones you define, like `Person`—must either use or override. The `Any` class is shown in its entirety in Listing 3.4.

> **NOTE** *It's easy to get class and object mixed up. A class is your code that defines an object. So,* `Person.kt` *is your* `Person` *class. The class defines an object. So, the actual code—which is a class—defines an object. Then, when you create a new instance of that object, you have an object instance. All three of these things are different: class, object, and instance.*
>
> *This difference is really important; more so because a lot of the time, developers are a little sloppy in terminology and use the terms interchangeably. Your ability to determine if "create a new* `Person` *object" means to create a new class, a new object, or a new object instance, will help you enormously in becoming a better developer.*

LISTING 3.4: Any, the base class for all Kotlin classes

```
/*

 * Copyright 2010-2015 JetBrains s.r.o.

 *

 * Licensed under the Apache License, Version 2.0 (the "License");

 * you may not use this file except in compliance with the License.

 * You may obtain a copy of the License at

 *

 * http://www.apache.org/licenses/LICENSE-2.0

 *
```

LISTING 3.4 *(continued)*

```
 * Unless required by applicable law or agreed to in writing, software

 * distributed under the License is distributed on an "AS IS" BASIS,

 * WITHOUT WARRANTIES OR CONDITIONS OF ANY KIND, either express or implied.

 * See the License for the specific language governing permissions and

 * limitations under the License.

 */

package kotlin

/**

 * The root of the Kotlin class hierarchy. Every Kotlin class has [Any] as a
superclass.

 */

public open class Any {

    /**

     * Indicates whether some other object is "equal to" this one. Implementations
must fulfil the following

     * requirements:

     *

     * * Reflexive: for any non-null value `x`, `x.equals(x)` should return true.

     * * Symmetric: for any non-null values `x` and `y`, `x.equals(y)` should
return true if and only if `y.equals(x)` returns true.

     * * Transitive:  for any non-null values `x`, `y`, and `z`, if `x.equals(y)`
returns true and `y.equals(z)` returns true, then `x.equals(z)` should return true.

     * * Consistent:  for any non-null values `x` and `y`, multiple invocations of
`x.equals(y)` consistently return true or consistently return false, provided no
information used in `equals` comparisons on the objects is modified.

     * * Never equal to null: for any non-null value `x`, `x.equals(null)` should
return false.

     *
```

```
    * Read more about [equality](https://kotlinlang.org/docs/reference/equality
.html) in Kotlin.

    */

    public open operator fun equals(other: Any?): Boolean


    /**

    * Returns a hash code value for the object.  The general contract of
`hashCode` is:

    *

    * * Whenever it is invoked on the same object more than once, the `hashCode`
method must consistently return the same integer, provided no information used in
`equals` comparisons on the object is modified.

    * * If two objects are equal according to the `equals()` method, then calling
the `hashCode` method on each of the two objects must produce the same inte-
ger result.

    */

    public open fun hashCode(): Int

    /**

    * Returns a string representation of the object.

    */

    public open fun toString(): String

}
```

Always Override hashCode() and equals(x)

Notice the last part of the definition of hashCode() from Listing 3.4:

```
    * * If two objects are equal according to the `equals()` method, then calling
the `hashCode` method on each of the two objects must produce the same integer
result.
```

So Kotlin makes it really clear—any two objects that are considered equal should return identical hash codes. But now that you've updated equals(x), is that the case?

It's easy to find out for yourself. Make the additions to your PersonTest.kt class so that it matches Listing 3.5. Some of these are new, to print out hash codes, but also note that the messages from equality checks have been made clearer.

LISTING 3.5: Updating the test class to allow for comparing equals results with hash code results

```kotlin
import org.wiley.kotlin.person.Person

fun main() {
    // Create a new person
    val brian = Person("Brian", "Truesby", 68.2, 33, true)
    println(brian)

    // Create another person
    val rose = Person("Rose", "Bushnell", 56.8, 32, true)
    println(rose)

    // Change Rose's last name
    rose.lastName = "Bushnell-Truesby"
    println(rose)

    if (rose.equals(brian))
        println("Rose and Brian are the same")
    else
        println("Rose and Brian aren't the same")

    val second_brian = Person("Brian", "Truesby", 68.2, 33, true)
    if (second_brian.equals(brian))
        println("Brian and Second Brian are the same!")
    else
        println("Brian and Second Brian are not the same!")

    val copy_of_brian = brian
    if (copy_of_brian.equals(brian))
        println("Brian and Copy of Brian are the same!")
    else
        println("Brian and Copy of Brian are not the same!")

    println("Brian hashcode: ${brian.hashCode()}")
    println("Rose hashcode: ${rose.hashCode()}")
    println("Second Brian hashcode: ${second_brian.hashCode()}")
    println("Copy of Brian hashcode: ${copy_of_brian.hashCode()}")
}
```

Compile and run this, and you'll get something similar to this:

```
Brian Truesby
Rose Bushnell
Rose Bushnell-Truesby
Rose and Brian aren't the same
Brian and Second Brian are the same!
Brian and Copy of Brian are the same!
Brian hashcode: 2083562754
Rose hashcode: 1239731077
Second Brian hashcode: 557041912
Copy of Brian hashcode: 2083562754
```

There's an immediate problem here: The `brian` instance and `second_brian` instance report as equals, but their hash codes don't match. And, potentially another surprise: `brian` and `copy_of_brian` are equal and their hash codes *do* match.

Default Hash Codes Are Based on Memory Location

The issue here is that, without you overriding `hashCode()`, it's going to report an integer based largely on the memory location of the object instance upon which it's called.

Practically, that means that two object instances with the same property values are *not* going to have the same hash code, because those two object instances live in different spaces within your system's memory. That's why `brian` and `second_brian` are returning different hash codes: they are two different pieces of memory, and therefore two different hash codes (and therefore a problem).

On the other hand, if you have two variables referencing the *same* object instance, then you're going to get identical hash codes. In fact, you're not getting two hash codes; you're getting the same hash code from the same object, but you're just calling that method twice. That's why the hash codes for `brian` and `copy_of_brian` are the same.

> **NOTE** *Don't get too hung up on exactly how the default implementation of* `hashCode()` *works. It's enough to know that it uses memory location, and that you're going to need to override it anytime you override* `equals(x)`.

So there's work to do. Instances that are the same (like `brian` and `copy_of_brian`) as well as instances that report back as equals using `equals(x)` (like `brian` and `second_brian`) should all return the same hash code.

Use Hash Codes to Come Up with Hash Codes

That heading sounds confusing, but it's actually a good bit of advice: you can use objects that have working `hashCode()` methods to build up your own hash code. Take `Person`; right now, the key properties—the ones that determine equality—are `firstName` and `lastName`. If you're not sure why, take a look again at how you overrode `equals(x)`:

```
override fun equals(other: Any?): Boolean {
    if (other is Person) {
        return (firstName.equals(other.firstName)) &&
                (lastName.equals(other.lastName))
    } else {
        return false
    }
}
```

Equality is basically controlled by those two properties; therefore, `hashCode()` should be as well. Better still, since those properties are just `String`s, you can rely on them having working `hashCode()` methods, as that's the job of the Kotlin engine you're using.

So you can use something like this:

```
override fun hashCode(): Int {
    return firstName.hashCode() + lastName.hashCode()
}
```

This code takes the two unique property names and uses the sum of their hash codes to get a unique value for the class instance overall. Compile this and run your test program and you should get output similar to this:

```
Brian hashcode: 680248194
Rose hashcode: -1533460291
Second Brian hashcode: 680248194
Copy of Brian hashcode: 680248194
```

> **NOTE** *Can you think of any problems with this implementation, aside from two people having the same first name and last name? There's a subtle but important problem that you'll have to come back and fix later.*

Notice that the two references to the same object instance *and* the instance with identical values for first and last name all now report the same hash code.

> **NOTE** *If you're wondering about that negative hash code for the* rose *object, it means that adding the two hash codes (from* firstName *and* lastName*) exceeded the upper limit of the* Int *class, and the value then wrapped around. This is an unusual behavior that you don't have to worry about too much for now.*

Unfortunately, you're not quite done yet.

SEARCHING (AND OTHER THINGS) DEPEND ON USEFUL AND FAST EQUALS(X) AND HASHCODE()

A couple of issues are worth examining before moving on:

➤ An object with the first name of another object's last name, and the last name of that other object's first name, are going to return the same hash code. That's bad.

➤ The current implementation of equals(x) could be faster; and with these methods, speed matters.

The first problem is not that hard to solve; you're just going to have to get a little creative, and also learn a bit about typical solutions to this sort of problem.

Multiple Properties to Differentiate Them in hashCode()

You need a way to see if this is happening (and when it's not). Add this to `PersonTest`:

```
val backward_brian = Person("Truesby", "Brian", 67.6, 42, false)
if (backward_brian.equals(brian))
    println("Brian and Backward Brian are the same!")
else
    println("Brian and Backward Brian are not the same!")

println("Brian hashcode: ${brian.hashCode()}")
println("Rose hashcode: ${rose.hashCode()}")
println("Second Brian hashcode: ${second_brian.hashCode()}")
println("Copy of Brian hashcode: ${copy_of_brian.hashCode()}")
println("Backward Brian hashcode: ${backward_brian.hashCode()}")
```

The output now should look similar to this:

```
Rose and Brian aren't the same
Brian and Second Brian are the same!
Brian and Copy of Brian are the same!
Brian and Backward Brian are not the same
Brian hashcode: 680248194
Rose hashcode: -1533460291
Second Brian hashcode: 680248194
Copy of Brian hashcode: 680248194
Backward Brian hashcode: 680248194
```

This definitely confirms the issue: `brian` and `backward_brian` are *not* equal, but they are returning the same hash code.

To fix this, you've got to differentiate between names matching, and specifically the first and last name matching. The way most developers handle this in a method like `hashCode()` is to use a multiplier that is different for each, like this:

```
override fun hashCode(): Int {
    return (firstName.hashCode() * 28) + (lastName.hashCode() * 31)
}
```

> *There is nothing significant about the number 28 or 31. You just need something to multiply the properties by. What is important is that you use a different number for each property. If you don't, you're going to still get identical hash codes when you shouldn't.*

Now, even if two objects have a match between first name and last name, and then last name and first name, the result from `hashCode()` will be different because the multiplier is different. Run `Person-Test` again to see this in action:

```
Rose and Brian aren't the same
Brian and Second Brian are the same!
Brian and Copy of Brian are the same!
```

```
Brian and Backward Brian are not the same
Brian hashcode: -580500212
Rose hashcode: -300288362
Second Brian hashcode: -580500212
Copy of Brian hashcode: -580500212
Backward Brian hashcode: 2060437994
```

Use == over equals(x) for Speed

The second item is to consider some speed improvements in equals(x). This is important because one of the key uses for equals(x) is in list comparison. You've not used lists yet, but they're an important part of programming languages, and Kotlin is no exception. On top of that, one key part of using lists is comparing items in them.

Whether you're building a web application or a mobile application, ensuring that every time two items are compared in a list is as fast as possible is a *big deal* for your users' experience.

An easy performance gain is to be had in looking at the current equals(x) implementation:

```kotlin
override fun equals(other: Any?): Boolean {
    if (other is Person) {
        return (firstName.equals(other.firstName)) &&
                (lastName.equals(other.lastName))
    } else {
        return false
    }
}
```

This is using equals(x) to do string comparisons. But that method actually uses == (the double equals operator), which compares two objects (Strings in this case) to see if they're equals. You can rewrite equals(x) to look like this:

```kotlin
override fun equals(other: Any?): Boolean {
    if (other is Person) {
        return (firstName == other.firstName) &&
                (lastName == other.lastName)
    } else {
        return false
    }
}
```

While this doesn't seem like a big deal, it's going to result in two fewer method calls: one to equals(x) on firstName and another on equals(x) in lastName. Those methods would then just use ==, so you can go directly to that. Two method calls on every comparison can really add up, especially in big lists full of hundreds of objects!

A Quick Speed Check on hashCode()

You should also take a look at hashCode() and see if it is also optimized for the fast comparisons that list operations need. Here's the current version:

```kotlin
override fun hashCode(): Int {
    return (firstName.hashCode() * 28) + (lastName.hashCode() * 31)
}
```

Fortunately, there's not much needed here, for a few reasons:

➤ You're using the hashCode() methods from the built-in String class. You can generally count on hashCode() implementations on Kotlin basic types being well-optimized, so that's a safe thing to leave in place.

➤ You're using multiplication and addition of Ints. This is another highly optimized activity that isn't going to take a lot of CPU time.

➤ Finally, you're adding the results of those multiplications. Addition is also blazing fast.

So this method is good to go. Still, when you check out equals(x), always remember to check out hashCode(), too.

BASIC CLASS METHODS ARE REALLY IMPORTANT

At this point, you may be a little surprised. After all, in over 15 pages of text on learning Kotlin, you've spent that entire time working with two methods: equals(x) and hashCode(). On top of that, these are methods that already were written for you! You've just been updating them. So what gives?

The reality is that the fundamentals of class design in Kotlin are really, really important. The very best programmers—in any language—understand that a well-written equals(x) method can dramatically improve performance of a program, reduce bugs related to checking for equality, and ultimately reflect a skilled hand at the class helm. Pick up classic programming works like Joshua Bloch's *Effective Java* or the famous Gang of Four's work on *Design Patterns: Elements of Reusable Object-Oriented Software* and you'll find a lot of time spent on basics just like this.

> **NOTE** *In fact, Joshua Bloch talks about the need to override* hashCode() *when you override* equals(x) *in* Effective Java *specifically.*

You've also implicitly taken some important steps to understanding inheritance, as well as gotten your first look at the Any class. These are both key to later chapters that dig much deeper into inheritance and subclassing. So take heart! All this work will pay off, and you'll not be among the masses left wondering "why do things work the way they do with Kotlin classes?"

Inheritance Matters

➤ Building more flexible classes with secondary constructors

➤ When do you need a subclass?

➤ All Kotlin classes inherit from `Any`

➤ Creating the option for inheritance

➤ Writing a superclass

GOOD CLASSES ARE NOT ALWAYS COMPLEX CLASSES

So far, you've largely focused on a single class: the `Person` begun in Chapter 1 and developed in the last few chapters. It's a good "first class" because it represents a real-world object and raises lots of core issues, related to equality, mutable properties, and construction.

It's possible to begin to let some concern creep in, though: are classes really this simple? Is this just a "toy" class that isn't realistic? Well, yes and no.

Yes, in that `Person` is a pretty typical class. It represents something, it has some properties, and it's kept intentionally basic. In other words, it has nothing that isn't absolutely essential.

No, in that some key things are missing from `Person`:

➤ It doesn't inherit from a custom base class, or get extended by any other classes.

➤ It doesn't offer multiple constructors.

➤ It doesn't offer any real behavior, just properties.

This chapter is going to address several of those "No" answers and give you a taste of "real-world" Kotlin classes along the way.

Keep It Simple, Stupid

Before getting into more details, it's worth restating the childhood expression: Keep it simple. In other words, don't add lots of properties and methods "just in case." Instead, just write the code you need.

To see this in action, Listing 4.1 shows the code for `CardViewActivity`, a class for using a piece of the Android 5.0 support library.

LISTING 4.1: CardViewActivity

```
/*
 * Copyright (C) 2017 The Android Open Source Project
 *
 * Licensed under the Apache License, Version 2.0 (the "License");
 * you may not use this file except in compliance with the License.
 * You may obtain a copy of the License at
 *
 *       http://www.apache.org/licenses/LICENSE-2.0
 *
 * Unless required by applicable law or agreed to in writing, software
 * distributed under the License is distributed on an "AS IS" BASIS,
 * WITHOUT WARRANTIES OR CONDITIONS OF ANY KIND, either express or implied.
 * See the License for the specific language governing permissions and
 * limitations under the License.
 */

package com.example.android.cardview

import android.app.Activity
import android.os.Bundle

/**
 * Launcher Activity for the CardView sample app.
 */
class CardViewActivity : Activity() {

    override fun onCreate(savedInstanceState: Bundle?) {
        super.onCreate(savedInstanceState)
        setContentView(R.layout.activity_card_view)
        if (savedInstanceState == null) {
            fragmentManager.beginTransaction()
                    .add(R.id.container, CardViewFragment())
                    .commit()
        }
    }
}
```

Notice how little is in this class! Obviously this is a somewhat extreme example, but it's still an example nonetheless: nothing is here that isn't needed. It simply extends a base class (more on this soon), details what should happen when the `Card` is created, and then has a little bit of behavior.

That's the idea: whenever possible, simple code, using inheritance, gets the job done better than thousands of lines of code.

> **NOTE** *Obviously, at times you are going to need a lot more complexity. But the point here is that those times are probably less than you think. Especially if you plan on writing mobile applications using Kotlin, you're going to want to keep things simple. The more code you write, the larger your program, and that is an issue even on today's hard-drive-sized phones.*

Keep It Flexible, Stupid

This is admittedly not a childhood saying—but maybe it should be. So far, you've been creating a simple `Person` class, and the main concern has been giving it some properties and then overriding `equals(x)` and `hashCode()`.

One thing that hasn't been discussed is really thinking through how this class will be used—both by programs and by other classes. That's another topic you're going to see throughout this chapter. Every decision you make about `Person` affects programs that use `Person`, and other objects that may extend `Person`.

To help you think about this as we get into the chapter, consider Listing 4.2, a very simple class called `Square`.

LISTING 4.2: Basic class to "draw" a square

```
open class Square {
    val length: Int = 12

    open fun drawSquare() { println("Drawing a square with area " +
                                    (length * length)) }
}
```

This class defines (shockingly!) how to draw a `Square`. No big deal, right?

> **NOTE** *You'll see some new pieces of syntax here that you don't need to worry about right now: the* open *keyword or the use of* get(). *You'll learn about these throughout the rest of this chapter. For now, just focus on the basic class and that* drawSquare() *method.*

But this class is not very flexible; in fact, it's totally specific to a square, because of the name of its single method: `drawSquare()`. Consider what happens when another class wants to extend this, as in Listing 4.3.

LISTING 4.3: Extending Square to become a rectangle

```
class Rectangle: Square() {
    val width: Int = 8

    fun drawRectangle() {
        println("Drawing a rectangle with area " + (length * width))
    }
}
```

`Rectangle` now is a subclass of `Square`, so it gets a length property from `Square` and then adds its own property, `width`. It also inherits `drawSquare()` from `Square`, and adds its own `drawRectangle()`.

The problem here isn't that obvious, but will become more so as you work through this chapter: you now have *two* draw methods—`drawSquare()` (which came from the base class) and the new `drawRectangle()` method. That's not simple, and it's a result of `Square` not being flexible.

`Square` isn't flexible because it used a name for a method—`drawSquare`—that didn't consider that other classes might want to extend the class and override that method. A better name would simply be "draw," as shown in Listing 4.4.

LISTING 4.4: Reworking Square to use a more flexible property name

```
open class Square {
    val length: Int = 12

    open fun draw() { println("Drawing a square with area " +
                              (length * length)) }
}
```

Now, `Rectangle` can extend `Square` and just override that `draw()` method, as shown in Listing 4.5.

LISTING 4.5: Now extending Square is much cleaner

```
class Rectangle: Square() {
    val width: Int = 8

    override fun draw() {
        println("Drawing a rectangle with area " + (length * width))
    }
}
```

Because `Square` was adjusted to be more flexible, `Rectangle` is now more sensible: it overrides `draw()`, and now there is just a single `draw()` method for both classes.

This is the sort of consideration to take into account throughout the rest of this chapter, and is particularly key if you ever open-source or share your code with anyone. They're going to want clear and flexible classes, just as you would.

CLASSES CAN DEFINE DEFAULT VALUES FOR PROPERTIES

Look at Listing 4.6, which contains the current version of `Person`.

LISTING 4.6: The latest version of Person works, but isn't flexible

```kotlin
package org.wiley.kotlin.person

class Person(_firstName: String, _lastName: String,
             _height: Double, _age: Int, _hasPartner: Boolean) {

    var firstName: String = _firstName
    var lastName: String = _lastName
    var height: Double = _height
    var age: Int = _age
    var hasPartner: Boolean = _hasPartner

    fun fullName(): String {
        return "$firstName $lastName"
    }

    override fun toString(): String {
        return fullName()
    }

    override fun hashCode(): Int {
        return (firstName.hashCode() * 28) + (lastName.hashCode() * 31)
    }

    override fun equals(other: Any?): Boolean {
        if (other is Person) {
            return (firstName == other.firstName) &&
                   (lastName == other.lastName)
        } else {
            return false
        }
    }
}
```

This class certainly has done the job, but is it flexible? One way to answer this question is to look at what it takes to create a new instance of the class in question. For `Person`, referring back to `Person-Test`, you'll need a line of code like this:

```kotlin
val brian = Person("Brian", "Truesby", 68.2, 33, true)
```

Is this good? Well, it certainly gets the core values needed for a `Person`. But it goes beyond that—are there cases where you might not have a person's weight? And is it really necessary to specify if a `Person` instance has a partner at object instance construction?

These are examples where the class just isn't that flexible. It requires more information than is always available.

Constructors Can Accept Default Values

To make this more flexible, you can use *default values* for non-essential properties. A person is really defined by a few core pieces of data; in this case, that is the first and last name of the person. The other details are really useful, but not required.

Default values allow you to provide values for those non-essential properties that will be used if nothing else is provided. Here's a look at how to change the constructor for `Person` to do just this:

```
class Person(_firstName: String, _lastName: String,
             _height: Double = 0.0, _age: Int = 0,
             _hasPartner: Boolean = false) {
```

Kotlin will now allow you to skip values for anything that has a default: the person's height, age, and if they have a partner. So you could create a new instance like this:

```
// Create another person
val rose = Person("Rose", "Bushnell")
```

In this instance, the rose variable will have a height of 0.0, an age of 0, and no partner. This cleans up the code rather significantly.

Kotlin Expects Arguments in Order

There are a couple of wrinkles here, though. Suppose you want to create a person with just their name—which is now possible—but also to indicate that they have a partner. You *cannot* simply add that argument to the constructor. In other words, the following code is illegal:

```
val rose = Person("Rose", "Bushnell", true)
```

The `hasPartner` property is 5th, not 3rd. If you choose to take advantage of default values, you cannot skip any values. So to indicate a partner, you'd need values for the arguments before that one:

```
val rose = Person("Rose", "Bushnell", 0.0, 0, true)
```

This works, but now you've lost some of that flexibility again. You could just set the property after object instance creation:

```
val rose = Person("Rose", "Bushnell")
rose.hasPartner = true
```

This is likely your best bet. It's clear and doesn't require any extra code.

You can continue to specify argument values *in order* and leave off default values—as long as they are at the end of your construction. So this is legal, if you know a person's first name, last name, and height, but not their age or partner status:

```
val rose = Person("Rose", "Bushnell", 62.5)
```

Specify Arguments by Name

There is a way around this limitation: you can pass arguments into a constructor by name. This is not used that often but can help you get around the order limitation. You just supply the name of the argument as defined in the class, an equals sign, and a value:

```
val rose = Person("Rose", "Bushnell", _hasPartner = true)
```

Kotlin will happily pass the value supplied to the named argument. In fact, you can do this with all your arguments if you really want:

```
val rose = Person(_firstName = "Rose", _lastName = "Bushnell",
                 _hasPartner = true)
```

Of course, this is a bit verbose, and you'll rarely see this sort of syntax. And the biggest drawback is it requires a calling program to know the names of the arguments in the class. Parameter names like _firstName are often chosen because they aren't ever seen by the calling program. This isn't a huge issue, but it's worth thinking about.

Change the Order of Arguments (If You Need)

It's easy to create an instance and then assign it additional properties. But there will be times when you find that you are frequently setting the same property after construction. For example, suppose that over half the time, you want an instance with a first name, last name, and partner status. Currently, that means you'll always need to do this:

```
val rose = Person("Rose", "Bushnell")
rose.hasPartner = true
```

That's because the hasPartner input is last. Or, you would need to keep entering the parameter name:

```
val rose = Person("Rose", "Bushnell", _hasPartner = true)
```

Neither are bad, or even real problems. However, if you find that the most common usage of your constructor involves a different set of parameters, there's a simple fix! Just change the order of your constructor's arguments:

```
class Person(_firstName: String, _lastName: String,
             _hasPartner: Boolean = false,
             _height: Double = 0.0, _age: Int = 0) {
```

That's a simple change, but a really easy one, and now this construction is legal:

```
val rose = Person("Rose", "Bushnell", true)
```

This isn't a huge change, but does make things a little more direct, less verbose, and avoids the need to keep typing in that parameter name _hasPartner.

> **WARNING** *If you make this change to the constructor for* Person, *and then update* PersonTest, *you may get an error that takes you a second to process: that error occurs because now any constructors that supply all inputs will fail. That's because now the third argument is no longer a* Double *(for* height*) but instead a* Boolean *(for* hasPartner*). Just change those constructions, too, and you'll be back in business.*

SECONDARY CONSTRUCTORS PROVIDE ADDITIONAL CONSTRUCTION OPTIONS

Suppose you have a case where you want to allow entirely different information passed into an object at creation. In other words, you're not just decreasing the number of arguments passed to the constructor, but changing one or more of them entirely.

To make this concrete, consider a `Person` that has a partner. Right now, your class only knows if the person has a partner, but nothing about that partner. Let's change `Person` to allow an actual additional `Person` instance to be passed in *as the partner*.

So you really want to construct a `Person` like this:

```
val rose = Person("Rose", "Bushnell", brian)
```

Here, the `rose` instance is created and another `Person` instance, `brian`, is passed in as well. For this, you need something entirely new: a secondary constructor.

Secondary Constructors Come Second

This is a case where you don't just want to add another argument to the constructor of `Person`; you actually want to pass in something totally different: another `Person` instance to represent the constructed instance's partner.

This is where a secondary constructor comes in. You literally just add an additional constructor, using the `constructor` keyword:

```
constructor(_firstName: String, _lastName: String,
            _partner: Person) {
    // Need code here soon
}
```

This new constructor still takes in a first and last name, and adds a new argument, `_partner`, which has to be a `Person` instance. Now, you could start by trying to compile this, but you're going to immediately get an error:

```
Error:(14, 5) Kotlin: Primary constructor call expected
```

The problem here is that Kotlin expects all secondary constructors (and you can have as many as you want) to call the primary constructor, like this:

```
constructor(_firstName: String, _lastName: String,
            _partner: Person) :
            this(_firstName, _lastName) {
    // Need code here soon
}
```

What's going on here isn't obvious unless you remember what `this` references—and that was only briefly mentioned earlier, so don't worry if you don't recall. `this` is a reference to the current instance of the class. So `this(_firstName, _lastName)` is calling the class's default constructor, and passing in the values the secondary constructor received.

So this code is actually equivalent to something like this:

```
// Create a new Person instance like:
// newPerson = Person("First", "Last")
// Then run the code in the secondary constructor, with access to a
//   Person instance representing the partner
```

Another way of following this code is to trace the sequence of code that is run:

1. The secondary constructor is called.

2. The secondary constructor calls the primary constructor.

3. The primary constructor runs.

4. Any `init` block associated with the primary constructor is run, as well as property assignments.

5. Control is returned to the secondary constructor.

6. Any code in the secondary constructor is run.

7. Any code associated with the secondary constructor (through an additional code block) is run.

So now the actual code for the secondary constructor needs to be put into place.

Secondary Constructors Can Assign Property Values

At this point, you need to be aware of what the values of all your instance properties are. Because the default constructor has run, and has only been supplied a first and last name, the default values for age, height, and partner status have been used. That means, for example, that `hasPartner` is false.

Of course, that's a problem, because using the second constructor involves passing in a new `Person` instance, which implies the new person has a partner. So that property can be reassigned a correct value:

```
constructor(_firstName: String, _lastName: String,
          _partner: Person) :
          this(_firstName, _lastName) {
    hasPartner = true
}
```

Now, things get a bit trickier: what about that `Person` instance being passed in? Your instinct might be to do something like this:

```
constructor(_firstName: String, _lastName: String,
          _partner: Person) :
          this(_firstName, _lastName) {
    hasPartner = true
    partner = _partner
}
```

That makes sense. Storing the instance is helpful. Then you'd need to create the property itself, right below the existing properties:

```
var firstName: String = _firstName
var lastName: String = _lastName
var height: Double = _height
var age: Int = _age
var hasPartner: Boolean = _hasPartner
var partner: Person
```

Listing 4.7 shows the current version of Person, just to make sure you're following along.

LISTING 4.7: Current version of Person (with some problems!)

```
package org.wiley.kotlin.person

class Person(_firstName: String, _lastName: String,
             _hasPartner: Boolean = false,
             _height: Double = 0.0, _age: Int = 0) {

    var firstName: String = _firstName
    var lastName: String = _lastName
    var height: Double = _height
    var age: Int = _age
    var hasPartner: Boolean = _hasPartner
    var partner: Person

    constructor(_firstName: String, _lastName: String,
                _partner: Person) :
                this(_firstName, _lastName) {
        hasPartner = true
        partner = _partner
    }

    fun fullName(): String {
        return "$firstName $lastName"
    }

    override fun toString(): String {
        return fullName()
    }

    override fun hashCode(): Int {
        return (firstName.hashCode() * 28) + (lastName.hashCode() * 31)
    }

    override fun equals(other: Any?): Boolean {
        if (other is Person) {
            return (firstName == other.firstName) &&
                    (lastName == other.lastName)
        } else {
            return false
        }
    }
}
```

Now, to see if there are any issues, build or compile your project. You'll get an error related to these changes:

```
Error:(12, 5) Kotlin: Property must be initialized or be abstract
```

What does this mean? Well, you can pretty quickly guess that the property in question is the new one, `partner`. And the warning is right: you're not initializing the property with a value, and it's not abstract. For now, the focus should be on initialization.

> **NOTE** *We'll talk more about abstract properties later, when we dig much further into inheritance and subclassing. For now, consider that it's not what you want in terms of the* partner *property.*

Remember that in Kotlin, properties declared in a class must be explicitly given a value. That's why you took all those properties in during the primary construction and assigned them:

```
class Person(_firstName: String, _lastName: String,
             _hasPartner: Boolean = false,
             _height: Double = 0.0, _age: Int = 0) {

    var firstName: String = _firstName
    var lastName: String = _lastName
    var height: Double = _height
    var age: Int = _age
    var hasPartner: Boolean = _hasPartner
```

It's also why you have to give your properties a default value if they're not passed into a constructor with a value defined.

Here, though, that's really not the situation. You don't want to require a `partner` instance because not every `Person` instance has one. In other words, this is a rare case where you want to accept null values.

You Can Assign null to a Property . . . Sometimes

If all that's needed is an assignment to `partner`, you might try something like this:

```
var firstName: String = _firstName
var lastName: String = _lastName
var height: Double = _height
var age: Int = _age
var hasPartner: Boolean = _hasPartner
var partner: Person = null
```

This seems to meet the requirements of Kotlin, right? You've assigned the property a value: `null`. But it's still going to throw off an error:

```
Error:(12, 27) Kotlin: Null can not be a value of a non-null type Person
```

The issue is Kotlin's resistance to unexpected `null` values. You might remember this was first mentioned in Chapter 3, in the discussion of `Any` and the `?` operator. Remember this declaration of the `equals(x)` method?

```
override fun equals(other: Any?): Boolean {
```

`Any` indicated that any object that was derived from `Any` (so a subclass) was allowed, and the trailing `?` meant that null values were also allowed. So you could read this as "accept a subclass of `Any` or a `null`."

You want something similar here. You want to say, "accept an instance of `Person` or a null," which allows a partner (or not). So the `?` operator is again the answer:

```
var partner: Person? = null
```

That one additional `?` tells Kotlin to allow null values, which is what is needed here. Listing 4.8 shows this change, along with a small update to the secondary constructor to update _hasPartner to indicate that if a `Person` instance was passed in, the person being created does indeed have a partner.

LISTING 4.8: Person with a working secondary constructor

```
package org.wiley.kotlin.person

class Person(_firstName: String, _lastName: String,
             _hasPartner: Boolean = false,
             _height: Double = 0.0, _age: Int = 0) {

    var firstName: String = _firstName
    var lastName: String = _lastName
    var height: Double = _height
    var age: Int = _age
    var hasPartner: Boolean = _hasPartner
    var partner: Person? = null

    constructor(_firstName: String, _lastName: String,
                _partner: Person) :
                this(_firstName, _lastName) {
        hasPartner = true
        partner = _partner
    }

    fun fullName(): String {
        return "$firstName $lastName"
    }

    override fun toString(): String {
        return fullName()
    }

    override fun hashCode(): Int {
        return (firstName.hashCode() * 28) + (lastName.hashCode() * 31)
    }

    override fun equals(other: Any?): Boolean {
```

```
        if (other is Person) {
            return (firstName == other.firstName) &&
                   (lastName == other.lastName)
        } else {
            return false
        }
    }
}
```

You can test this out with Listing 4.9, a simplified `PersonTest` for exercising this new constructor.

LISTING 4.9: Testing out the secondary constructor of Person

```
import org.wiley.kotlin.person.Person

fun main() {
    // Create a new person
    val brian = Person("Brian", "Truesby", true, 68.2, 33)
    println(brian)

    // Create another person
    val rose = Person("Rose", "Bushnell", brian)
    println(rose)

    if (rose.hasPartner) {
        println("$rose has a partner: ${rose.partner}.")
    } else {
        println("$rose does not have a partner.")
    }

    // Change Rose's last name
    rose.lastName = "Bushnell-Truesby"
    println(rose)
}
```

Run this and you'll see the following output:

```
Brian Truesby
Rose Bushnell
Rose Bushnell has a partner: Brian Truesby.
Rose Bushnell-Truesby
```

This is exactly what is desired: the secondary constructor works and captures both the partner status of the new `Person` instance as well as a reference to the partner instance.

null Properties Can Cause Problems

It's important to realize what you've done, and what you haven't. Here's a brief summary of how null relates to the work you have in `Person`:

➤   You cannot pass `null` for any values in the `Person` primary constructor.

➤   You can assign the partner property a `null` value.

➤ You cannot pass a `null` value into the secondary `Person` constructor.

➤ You can directly set the partner property to `null`.

This actually makes for an interesting combination of events. Update `PersonTest` to look like this:

```
// Create another person
val rose = Person("Rose", "Bushnell", brian)
println(rose)
rose.partner = null

if (rose.hasPartner) {
    println("$rose has a partner: ${rose.partner}.")
} else {
    println("$rose does not have a partner.")
}
```

See any issues? The output will show you a problem:

```
Rose Bushnell has a partner: null.
```

This happens because you're allowing `partner` to directly be set to `null`, but not updating `hasPartner`. That's got to be fixed.

> **WARNING** *Unfortunately, when you start allowing* `null` *values, you're going to often run into this sort of issue. Kotlin tries very hard to prevent you from using* `null`. *When you use* `null` *anyway—and there will be times where it makes sense—you're going to have to check your code a few times for issues just like this.*

HANDLE DEPENDENT VALUES WITH CUSTOM MUTATORS

The issue is that you now have two properties that are closely related: `hasPartner` and `partner`. Right now, those values are separate. You can set them separately, and in fact you can pass in `hasPartner` to the primary constructor and `partner` to the secondary constructor.

There's some work to do, and it is common work; you'll often find a class (or classes) with dependent values that need to interoperate.

Set Dependent Values in a Custom Mutator

This is a perfect case for using your own custom mutator (or setter, if you like). You saw custom mutators back in Chapter 2, although ultimately you didn't need one. Here, though, you do.

Whenever a partner is set, the `hasPartner` property should also be set. Simple enough:

```
var partner: Person? = null
    set(value) {
        field = value
        hasPartner = true
    }
```

> **NOTE** *If you're hazy on the details regarding field and value, flip back to Chapter 2 where overriding the default mutator was first detailed. Re-read that section and then come back here for more.*

This means that code like this:

```
rose.partner = brian
```

is now essentially the same as this:

```
rose.partner = brian
rose.hasPartner = true
```

This is good; it ensures that when a `Person` instance has a partner, they will be set to correctly report back on `hasPartner`. But you now have some subtle issues that still need cleanup.

All Property Assignments Use the Property's Mutator

Take a look at the current secondary constructor for `Person`:

```
constructor(_firstName: String, _lastName: String,
            _partner: Person) :
            this(_firstName, _lastName) {
    hasPartner = true
    partner = _partner
}
```

When you assign _partner to partner, you can think of that code as being "replaced" by the mutator code. So this code really becomes:

```
constructor(_firstName: String, _lastName: String,
            _partner: Person) :
            this(_firstName, _lastName) {
    hasPartner = true
    // This:
    //    partner = _partner
    // gets expanded into this:
    partner = _partner // field = value
    hasPartner = true

}
```

See anything interesting here? You should. `hasPartner` is actually set *twice*. Your custom mutator is now handling that assignment, so you can remove it from your secondary constructor:

```
constructor(_firstName: String, _lastName: String,
            _partner: Person) :
            this(_firstName, _lastName) {
    partner = _partner
}
```

This is much cleaner now.

Nullable Values Can Be Set to null!

There's another issue, though, and it's related to this property declaration:

```
var partner: Person? = null
```

The issue is that because `partner` is allowed to take on the value of `null`, `partner` can actually be set to `null` explicitly. You even did this earlier:

```
// Create another person
val rose = Person("Rose", "Bushnell", brian)
println(rose)
rose.partner = null

if (rose.hasPartner) {
    println("$rose has a partner: ${rose.partner}.")
} else {
    println("$rose does not have a partner.")
}
```

Setting `rose.partner` to `null` is a perfectly legal assignment. But now run this code, and you're going to get incorrect output:

```
Rose Bushnell has a partner: null.
```

What's going on here? Well, your mutator for `partner` is setting `hasPartner` to true in every case. It should actually be checking the passed-in `partner` (which might be `null`!) and then making a decision about what to set `hasPartner` to based on that check.

You can check for `null` in Kotlin using the `==` operator, like this:

```
if (value == null) {
    println("It's null!")
}
```

Similarly, you can check that a value is *not* `null` with `!=`, like this:

```
if (value != null) {
    println("It's not null!")
}
```

So now you can rewrite the mutator for `partner` to check for `null`, and then set `hasPartner` appropriately:

```
var partner: Person? = null
    set(value) {
        field = value
        if (field == null) {
            hasPartner = false
        } else {
            hasPartner = true
        }
    }
```

This code is perfectly workable. However, before moving on, you can actually make this a little simpler. In Kotlin, anytime you have an expression, that expression itself evaluates to a `Boolean`: it's either true or false.

So the following expression:

```
(field == null)
```

is going to return true if `field` is null, and false if it's not. That's almost what we want. If `field` (which represents the `partner` property in this case) is null, we want `hasPartner` to be false; if field is *not* null, `hasPartner` should be true.

In other words, `hasPartner` should be set to the evaluation of the expression `(field != null)`.

~~~~~~~~~~~ your secondary constructor now:

```
                                   null)

                              ing 4.10 shows the latest version of Person.
```

___ secondary constructor and a custom mutator

```
on

      ing, _lastName: String,
      blean = false,
      = 0.0, _age: Int = 0) {

      _firstName
      lastName
      eight

      = _hasPartner
      null

      field != null)

      : String, _lastName: String,
      Person) :
            tName, _lastName) {
    partner = _partner
}

fun fullName(): String {
    return "$firstName $lastName"
}

override fun toString(): String {
    return fullName()
}
```

continues

LISTING 4.10 *(continued)*

```kotlin
    override fun hashCode(): Int {
        return (firstName.hashCode() * 28) + (lastName.hashCode() * 31)
    }

    override fun equals(other: Any?): Boolean {
        if (other is Person) {
            return (firstName == other.firstName) &&
                   (lastName == other.lastName)
        } else {
            return false
        }
    }
}
```

At this point, you should test this code for yourself. Experiment with changing `PersonTest` to add a person with a partner, remove the person's partner, and check if they have a partner (using `hasPartner`). Add a new partner, or add a null partner. All of these operations should work, and `hasPartner` should reflect the person's correct status.

Limit Access to Dependent Values

At this point, things are coming along. But there's still the possibility for misuse. Suppose you put this code in `PersonTest`:

```kotlin
// Create another person
val rose = Person("Rose", "Bushnell", brian)
println(rose)
rose.hasPartner = false

if (rose.hasPartner) {
    println("$rose has a partner: ${rose.partner}.")
} else {
    println("$rose does not have a partner.")
}
```

This isn't a good thing, as it's breaking the connection between the `partner` instance and the `hasPartner` property. You just fixed that in the custom mutator, so now you need to make sure that nobody can set the status of `hasPartner` directly. It should be completely based upon the presence (or lack) of a `partner` instance.

Fortunately, Kotlin makes this pretty easy. You can just add the word `private` before the accessor (`get`) or mutator (`set`) you want to prevent access to. In this case, it's just the mutator that is the issue:

```kotlin
var hasPartner: Boolean = _hasPartner
    private set
```

Now, try to compile your program, and you'll get an error if you have code like this:

```kotlin
// Create another person
val rose = Person("Rose", "Bushnell", brian)
```

```
println(rose)
rose.hasPartner = false
```

Kotlin reports that the setter (mutator) is private:

```
Error:(11, 5) Kotlin: Cannot assign to 'hasPartner': the setter
                is private in 'Person'
```

This is generally good practice, because hasPartner is a dependent value. It is always going to report based on the partner instance. So now you've prevented code from setting its value directly.

When Possible, Calculate Dependent Values

There's actually yet another step you can take to clean up Person. Given that hasPartner is always dependent on partner, there's really no need to have a hasPartner property on its own. Instead, you can represent this as a new method on Person, and simply check the status of partner every time the method is called.

Here's what that method in Person should look like. Go ahead and add this in now:

```
fun hasPartner(): Boolean {
    return (partner != null)
}
```

The only real change between how you'd use this method and the now-defunct hasPartner property is in the trailing parentheses that are required for a method. So you used to write this:

```
if (rose.hasPartner) {
    // do something
}
```

Now you'd write this:

```
if (rose.hasPartner()) {
    // do something
}
```

If you try to compile your code, you're going to have to chase down a few errors you've created, like the use of parentheses. Other than that, you mostly need to remove code. Listing 4.11 shows Person with all the code you should remove commented out. You can comment these lines out, but a better idea would be to remove them altogether.

LISTING 4.11: Update Person to have a new method, hasPartner(), and remove the property of the same name

```
package org.wiley.kotlin.person

class Person(_firstName: String, _lastName: String,
            // _hasPartner: Boolean = false,
            _height: Double = 0.0, _age: Int = 0) {

    var firstName: String = _firstName
    var lastName: String = _lastName
    var height: Double = _height
```

LISTING 4.11 *(continued)*

```kotlin
    var age: Int = _age
    //var hasPartner: Boolean = _hasPartner
    //    private set
    var partner: Person? = null
        // set(value) {
            // field = value
            //hasPartner = (field != null)
        // }

    constructor(_firstName: String, _lastName: String,
                _partner: Person) :
                this(_firstName, _lastName) {
        partner = _partner
    }

    fun fullName(): String {
        return "$firstName $lastName"
    }

    fun hasPartner(): Boolean {
        return (partner != null)
    }

    override fun toString(): String {
        return fullName()
    }

    override fun hashCode(): Int {
        return (firstName.hashCode() * 28) + (lastName.hashCode() * 31)
    }

    override fun equals(other: Any?): Boolean {
        if (other is Person) {
            return (firstName == other.firstName) &&
                   (lastName == other.lastName)
        } else {
            return false
        }
    }
}
```

Also note that with this change, the custom mutator for `partner` can go away as well. Finally, you can remove the `_hasPartner` argument from the primary constructor. Much cleaner code!

You Can Avoid Parentheses with a Read-Only Property

Now that you've updated all of those `hasPartner` calls to `hasPartner()`, here's a bit of a tip: you can actually avoid that altogether. Instead of writing a `hasPartner()` function, you could just create a read-only property with the same custom accessor (getter). Add this declaration right under the other properties defined right after the primary constructor:

```kotlin
val hasPartner: Boolean
    get() = (partner != null)
```

You can also remove the `hasPartner()` code altogether. Listing 4.12 shows the completed version of `Person`.

LISTING 4.12: Moving from a function back to a property, but this time, a read-only one

```kotlin
package org.wiley.kotlin.person
open class Person(_firstName: String, _lastName: String,
                  _height: Double = 0.0, _age: Int = 0) {

    var firstName: String = _firstName
    var lastName: String = _lastName
    var height: Double = _height
    var age: Int = _age
    val hasPartner: Boolean
      get() = (partner != null)

    var partner: Person? = null

    constructor(_firstName: String, _lastName: String,
                _partner: Person) :
                this(_firstName, _lastName) {
        partner = _partner
    }

    fun fullName(): String {
        return "$firstName $lastName"
    }

    fun hasPartner(): Boolean {
        return (partner != null)
    }

    override fun toString(): String {
        return fullName()
    }

    override fun hashCode(): Int {
        return (firstName.hashCode() * 28) + (lastName.hashCode() * 31)
    }

    override fun equals(other: Any?): Boolean {
        if (other is Person) {
            return (firstName == other.firstName) &&
                   (lastName == other.lastName)
        } else {
            return false
        }
    }
}
```

Is this better than using a `hasPartner()` function? Not particularly; it's functionally equivalent. However, it's probably a little more *idiomatic*. In other words, Listing 4.12 is a little more how an experienced Kotlin programmer would write the `Person` class.

> **NOTE** *This, like other parts of this book, can be a bit annoying. First, you add a* hasPartner *property. Then you change that to a* hasPartner() *function and add parentheses. Now,* hasPartner *has changed back to a property—read-only this time—and you need to remove those parentheses that you just added.*
>
> *As much as this can seem a bit like we are going back and forth, it is an attempt to demonstrate how development often happens: iteratively, changing one thing and then changing another, and doing work that sometimes seemingly gets thrown away. While it would be easy to just give you the "final" version of this code right away, it's much more interesting and effective to have you walk through these steps and see how this process would go in an actual development environment.*

Finally, you can go back and clean up PersonTest to remove any incorrect usage of constructors that have been changed. Listing 4.13 shows a sample of what a test class might look like.

LISTING 4.13: The PersonTest class has a helper function and works with the updated Person class

```kotlin
import org.wiley.kotlin.person.Person

fun main() {
    // Create a new person
    val brian = Person("Brian", "Truesby", 68.2, 33)
    printPersonPartnerStatus(brian)

    // Create another person
    val rose = Person("Rose", "Bushnell", brian)
    printPersonPartnerStatus(rose)

    // Change Rose's last name
    rose.lastName = "Bushnell-Truesby"
    println(rose)
}

fun printPersonPartnerStatus(person: Person) {
    if (person.hasPartner) {
        println("$person has a partner: ${person.partner}.")
    } else {
        println("$person does not have a partner.")
    }
}
```

Notice that a new helper function has been introduced: printPersonPartnerStatus(). This just takes some code that you might use multiple times and ensures that it only has to be written once. That's the DRY principle discussed earlier: Don't Repeat Yourself.

> **NOTE** *Once again, you've done a lot of work just to then remove a lot of the code you just wrote. Still, this is the way of a good programmer. Code, refactor, and eliminate waste. While you could certainly just copy a final state for* Person *way back in Chapter 1, you'd know a lot less about how classes work, and that's really the point, isn't it? So keep working through the examples—even when it means removing code you worked hard to write—and realize that nobody judges a programmer on the length of their code. In fact, most good programmers write less code than less experienced ones!*

NEED SPECIFICS? CONSIDER A SUBCLASS

Person is looking pretty solid at this point. But it's still a single class, and a relatively simple one at that. Suppose you wanted to build out a genealogy and model the various relationships. You might have a person with a partner (which is already handled), but you also want to model parents and children (and their children).

This is a case where you have a class—Person—but now want a more specific version of that class. That's important to understand: by adding things to Person, like a parent or children, you're making it less flexible; you're making the class apply to a subset of the original thing.

Put another way, all children are people (Persons, if you like), but not all Persons are children. Person is more general, and a child is more specific. If you're moving from the general to the specific, you should consider using a subclass rather than just adding more properties to your existing class.

Any Is the Base Class for Everything in Kotlin

You've already seen this in action when you overrode equals(x) and hashCode() and looked (briefly) at the source code for Any. It's shown again in Listing 4.14, although most of the comments have been stripped out for space.

LISTING 4.14: Any, the base class for all Kotlin classes

```
/*
 * Copyright 2010-2015 JetBrains s.r.o.
 *
 * Licensed under the Apache License, Version 2.0 (the "License");
 * you may not use this file except in compliance with the License.
 * You may obtain a copy of the License at
 *
 * http://www.apache.org/licenses/LICENSE-2.0
 *
 * Unless required by applicable law or agreed to in writing, software
 * distributed under the License is distributed on an "AS IS" BASIS,
 * WITHOUT WARRANTIES OR CONDITIONS OF ANY KIND, either express or implied.
```

continues

LISTING 4.14 *(continued)*

```
 * See the License for the specific language governing permissions and
 * limitations under the License.
 */

package kotlin

public open class Any {
    public open operator fun equals(other: Any?): Boolean

    public open fun hashCode(): Int

    public open fun toString(): String
}
```

This is the base class for *all* classes in Kotlin. So `Person`, even though you may not have realized it, and certainly didn't have to do anything, is a subclass of `Any`.

This is the ultimate example of taking a nonspecific class—`Any`—and making it more specific by extending it in `Person`. All `Person` instances are also `Any` instances, but all instances of `Any` are certainly not `Person` instances.

When you have a base class, any method defined that has the word `open` in its method definition can be used by and extended by the subclass. Take a look at the three methods in `Any`:

```
public open operator fun equals(other: Any?): Boolean

public open fun hashCode(): Int

public open fun toString(): String
```

> **NOTE** *For now, don't worry about the* operator *keyword. You'll learn about that soon, but for now, focus on* open.

All three of these are `open` for overriding, and of course that's exactly what's been done in `Person`:

```
override fun toString(): String { ... }

override fun hashCode(): Int { ... }

override fun equals(other: Any?): Boolean { ... }
```

So there's a pairing here: the base class uses `open` and the subclass then uses `override`.

{ . . . } Is Shorthand for Collapsed Code

One quick technical note that applies to this book, but also to most Kotlin IDEs and interacting with the Kotlin community online, is that you'll sometimes see notation like this:

```
override fun hashCode(): Int { ... }
```

This was used earlier in referencing the three methods in Person that overrode methods defined in Any. The open and close brackets with three dots simply means "there is code here, but it's been collapsed (or hidden)."

This is not actual Kotlin syntax. You couldn't type that into your editor and get a successful compilation. However, it makes showing code (online or in print) shorter if the details of the method don't actually matter for what's being discussed.

You'll also see this in your IDE as an option. Figure 4.1 shows the IntelliJ IDEA with code as it normally appears, and Figure 4.2 shows that same code collapsed. You can toggle back and forth by clicking the small + and − icons just to the left of each line that defines a block of code (see the closeup in Figure 4.3).

```
Person.kt ×    PersonTest.kt ×
20              return "$firstName $lastName"
21          }
22
23      fun hasPartner(): Boolean {
24          return (partner != null)
25      }
26
27      override fun toString(): String {
28          return fullName()
29      }
30
31      override fun hashCode(): Int {
32          return (firstName.hashCode() * 28) + (lastName.hashCode() * 31)
33      }
34
35      override fun equals(other: Any?): Boolean {
36          if (other is Person) {
37              return (firstName == other.firstName) &&
38                      (lastName == other.lastName)
39          } else {
40              return false
41          }
42      }
43  }
```

FIGURE 4.1: Code from Person expanded completely

```
 12
 13        constructor(_firstName: String, _lastName: String,
 14                _partner: Person) :
 15                this(_firstName, _lastName) {
 16            partner = _partner
 17        }
 18
 19        fun fullName(): String {
 20            return "$firstName $lastName"
 21        }
 22
 23        fun hasPartner(): Boolean {
 24            return (partner != null)
 25        }
 26
 27        override fun toString(): String {...}
 30
 31        override fun hashCode(): Int {...}
 34
 35        override fun equals(other: Any?): Boolean {...}
 43    }
```

FIGURE 4.2: You can collapse code blocks to the { . . . } form in your IDE

```
 27        override fun toString(): String {...}
 30
 31        override fun hashCode(): Int {...}
 34
 35        override fun equals(other: Any?): Boolean {...}
 43    }
```

FIGURE 4.3: Toggle between expanded and collapsed with the + and – icons

A Class Must Be Open for Subclassing

You should also notice that Any uses the keyword open at the very top level:

```
public open class Any { ... }
```

This is what allows subclassing to happen. Any is open for subclassing, and then the methods that use the open keyword are open for overriding. Anything that does *not* declare itself as open cannot be subclassed or overridden.

To see this in action, go ahead and create a new class. Call it Parent. This will be a specific subclass of Person that has one or more children. Listing 4.15 shows the very simple beginnings of this new class, also in the org.wiley.person package.

LISTING 4.15: Parent, an eventual subclass of Person

```
package org.wiley.kotlin.person

class Parent {
}
```

Now, to inherit from a class, you simply append a colon (`:`) and then the name of the base class when declaring the new class:

```
class Parent : Person {
}
```

Go ahead and try to compile, and you're going to see a couple of immediate errors:

```
Error:(3, 16) Kotlin: This type is final, so it cannot be inherited from
Error:(3, 16) Kotlin: This type has a constructor, and thus must be
initialized here
```

The first of these illustrates the `open` requirement. `Person` is not currently `open`, so is considered *final*, which means it cannot be subclassed.

Go back to `Person` and add `open` before the `class` keyword:

```
open class Person(_firstName: String, _lastName: String,
                  _height: Double = 0.0, _age: Int = 0) {
```

Now compile again, and you'll see that the first error related to `Person` being final has gone away:

```
Error:(3, 16) Kotlin: This type has a constructor, and thus must be ini-
tialized here
```

Now you need to fix this second error, which has to do with construction and initialization.

Terminology: Subclass, Inherit, Base Class, and More

Before getting to that error, it's worth taking a moment to get your head around some key terms that are going to come up a lot. These all enter the programmer's vocabulary when you get into inheritance and subclasses:

➤ *Inheritance*: The principle of one class inheriting from another, usually properties and methods.

➤ *Base class*: The class that is inherited from. `Any` is the base class for all Kotlin classes, and `Person` is the base class for `Parent`.

➤ *Superclass*: Another word for base class. This is in relation to another class. So `Person` is the superclass of `Parent`.

➤ *Subclass*: A class that inherits from another class. So `Parent` is a subclass of `Person`.

➤ *Extend*: You'll sometimes hear a subclass described as extending a base class. This is common, and something you should be aware of.

You don't need to necessarily memorize these; there's no quiz coming. However, they are common, and you'll likely pick up on most of these as you work through this book and advance in your own programming career.

These are also words that aren't specific to Kotlin, but apply to any language that supports inheritance.

A Subclass Must Follow Its Superclass's Rules

Now, back to that error:

```
Error:(3, 16) Kotlin: This type has a constructor, and thus must be
initialized here
```

The key to understanding this is to remember what it takes to create a new `Person` instance:

```
val brian = Person("Brian", "Truesby", 68.2, 33)
```

`Person` has a primary constructor, which means that it requires certain things to be instantiated. That behavior is then inherited by its subclass.

You need to define a primary constructor for `Parent` that takes in any required information and passes it on to `Person`, like this:

```
class Parent(_firstName: String, _lastName: String,
            _height: Double = 0.0, _age: Int = 0) :
            Person(_firstName, _lastName, _height, _age) {
}
```

You should be able to reason through what's happening here pretty easily. `Parent` takes in all the same arguments that `Person` does, and then passes them to the `Person` constructor.

Keep in mind, though, that the rule here is not that you must initialize a subclass in the same way that you initialize that class's superclass; it's only that you must initialize the superclass in a valid manner. So you could just as easily do this:

```
class Parent(_firstName: String, _lastName: String) :
            Person(_firstName, _lastName, 68.5, 34) {
}
```

This doesn't make much sense here (there's no reason to assume a `Parent` is always 68.5 inches tall and 34 years old), but it's still following the rules of the superclass.

A Subclass Gets Behavior from All of Its Superclasses

`Parent` is also going to have an `equals(x)` method, a `hashCode()` method, and a `toString()` method, because it inherits all those methods from its parent, `Person`.

More specifically, a subclass also gets all the methods its superclass has. Note that the word here is *has*, not *defines*. In other words, imagine that `Person` did not override `toString()`. It would still have access to that method, because `Person` inherits from `Any`, which defines the method.

> **NOTE** *To be a little more specific, your Kotlin installation provides an implementation of* `Any` *with behavior for* `toString()`, *and* `Person` *gets that implementation. You'll learn more about interfaces and implementation in the coming chapters.*

In this imaginary case, `Person` defines behavior for `hashCode()` and `equals(x)` but inherits behavior for `toString()`. `Parent` then gets *all* of that behavior, regardless of whether it is inherited directly from `Person` or from a `Person` superclass. In fact, you could have four or five layers of inheritance and the lowest subclass gets everything from all of its superclasses, regardless of how far "up" the inheritance chain they are.

YOUR SUBCLASS SHOULD BE DIFFERENT THAN YOUR SUPERCLASS

This may seem like an obvious statement, but it bears repeating: your subclass, which is a more specific thing, should be different than your superclass, which is a more general thing.

For this reason, you may want to be a bit suspicious of a subclass that just calls its superclass's constructor as-is. It may indicate that you're not really thinking about what makes your subclass unique.

In the case of `Parent`, what is it that makes a `Parent` unique and more specific than a `Person`? Well, it's that a `Parent` has a child or children, of course.

Subclass Constructors Often Add Arguments

If the existence of a child is the very thing that differentiates a `Parent` from a `Person`, then consider making it required upon instantiation. Consider taking in the child (or children, something to address shortly) as part of the constructor.

Right now, there's no `Child` class (something else to get into a bit later). But a child is really just another `Person` instance, much as a partner was a `Person` instance. So you could add a parameter for this to your `Parent` constructor:

```
class Parent(_firstName: String, _lastName: String,
             _height: Double = 0.0, _age: Int = 0, _child: Person) :
             Person(_firstName, _lastName, _height, _age) {

    var child: Person = _child
}
```

This should intuitively feel right. You're now *requiring* the thing that defines a `Parent` as part of its construction. You certainly *could* just make this a property that can be set later, but then you could theoretically have a `Parent` that has no children, which is odd and possibly broken in this scenario.

> **NOTE** *At this point, it should be made clear: valid scenarios exist where an actual parent does not have a child. Many of those reasons are tragic, and this example isn't meant to be insensitive or invalidate those parents. It's instead a simplification to teach programming, and it actually illustrates a quite important principle: no program can ever truly model all the nuances and brilliance of real life.*

Now you have a class that does better model a `Parent`: it requires a child (via `Person` instance) to be passed in at construction.

Also take note that child is *not* a nullable property. It does not have the ? operator:

```
var child: Person? = _child
```

That's because this property is essential for a `Parent` and should never be null.

Don't Make Mutable What Isn't Mutable

It's at this point that you need to think about an important detail. Right now, your `child` property is mutable, because it's declared as a `var`:

```
var child: Person? = _child
```

But is that accurate? Should you be able to change out a child? That doesn't really model the real world. Listing 4.16 shows an updated test class.

LISTING 4.16: Updated test class

```
import org.wiley.kotlin.person.Person
import org.wiley.kotlin.person.Parent

fun main() {
    // Create a new person
    val brian = Person("Brian", "Truesby", 68.2, 33)
    printPersonPartnerStatus(brian)

    // Create another person
    val rose = Person("Rose", "Bushnell", brian)
    printPersonPartnerStatus(rose)

    // Change Rose's last name
    rose.lastName = "Bushnell-Truesby"
    println(rose)

    val mom = Parent("Cheryl", "Truesby", _child=brian)
    mom.child = rose
}

fun printPersonPartnerStatus(person: Person) { ... }
```

> **WARNING** *Don't forget to add the new import line at the top to import the* `Parent` *class in addition to* `Person`.

This is weird, to say the least. So it would be better to change child to a `val`, making it immutable:

```
val child: Person? = _child
```

Sometimes Objects Don't Exactly Map to the Real World

At this point, it's worth saying that while the goal in programming, and especially object-oriented programming, is to align with the real-world object your programmatic constructs are modeling, that doesn't always make sense.

In this case, it may be that setting a child through a constructor and never allowing it to change isn't what you want. That does make changing a child value difficult; you'd have to throw away the `Parent` instance and create a new one. That becomes a judgment call rather than an issue of right or wrong. You'll need to decide how your objects are used in your code, and also how they might be used by other people in their own code.

Remember, this chapter began with the theme of flexibility, and here's a case where you need to make some decisions around that flexibility. Do you want to model the case where a child really can't be "changed" on a `Parent` instance, which more closely models a real-world situation? Or do you want to use `var` instead, and allow change to make the object more convenient?

Generally, Objects Should Map to the Real World

In this case, you can leave `child` as a `val`. It may need to change later, and that's easy, but if you default to modeling real-world behavior, you'll tend to write better code and objects.

There's another real-world consideration to take into account now, as well: a parent can have more than one child. In fact, parents can have one or more children, and that's going to require some new concepts: collections, lists, and sets. That's what the next chapter is all about, so once you've got your head around the basics of inheritance, turn the page, and you can get into handling multiple children.

5

Lists and Sets and Maps, Oh My!

WHAT'S IN THIS CHAPTER?

➤ The basic `Collection` class in Kotlin

➤ Lists and how to use them

➤ Sets and how they differ from lists

➤ Iteration and mutability up close(r)

➤ One more collection: maps

LISTS ARE JUST A COLLECTION OF THINGS

At this point, it's time to take a detour back into some of the basic types that Kotlin provides. Currently, you can provide a `Parent` a single child instance through a `Person`. But parents can have multiple children (rather obviously). In fact, in programming terms, a parent can have a list of children: more than one. Kotlin supports lists in a variety of fashions.

Kotlin Lists: One Type of Collection

A Kotlin list is a type of the base Kotlin type for groupings of things: a `Collection`. `Collection`, part of the `kotlin.collections` package, defines a number of collection basics as well as ways to iterate (or traverse) over collections.

`Collections` defines methods to create various types of lists (and sets and maps, which are variations on a list), and those lists come in two basic varieties:

➤ *Read-only*: You can access elements in the collection, but not change them.

➤ *Mutable*: You can not only access elements but also add, remove, and update those elements.

Like you've already seen with much of Kotlin, you'll need to make decisions on the type of collection you want to work with at creation time. And most of the time when you need collections, you'll begin with lists. Listing 5.1 shows a super simple test program called `CollectionTest.kt` that you can use to start experimenting with this chapter.

LISTING 5.1: Beginning to play with lists

```
fun main() {
    val rockBands = mutableListOf("The Black Crowes", "Led Zeppelin",
                                    "The Beatles")
    println(rockBands)
}
```

This does just what it looks like: creates a mutable (changeable) list of things. In this case, the list holds strings. Then, `println()` uses the list's built-in `toString()` method to output the following:

```
[The Black Crowes, Led Zeppelin, The Beatles]
```

You can see that by default, a list prints out its members separated by commas. You'll also notice what will become a very familiar pattern in Kotlin when working with collections:

1. Create the collection

2. Use the collection

Those steps seem obvious, but it's important to take a minute to see that the creation of a list (or set or map) in Kotlin doesn't work quite like you might expect.

Collection Is a Factory for Collection Objects

`Collection` uses a loose form of the *factory pattern*, a common design pattern written about in depth in *Design Patterns: Elements of Reusable Object-Oriented Software*, already referenced earlier in this book. It's a pattern that describes a creation of objects (like a collection) *without* the use of the new keyword or object instantiation.

You've been creating new `Person` instances like this:

```
val brian = Person("Brian", "Truesby", 68.2, 33)
```

But you didn't do something similar for creating a new list. You did *not* do this:

```
val rockBands = List("The Black Crowes", "Led Zeppelin",
                        "The Beatles")
```

Instead, you used `mutableListOf()`:

```
val rockBands = mutableListOf("The Black Crowes", "Led Zeppelin",
                                "The Beatles")
```

In fact, `mutableListOf()` is a method on `Collection`, not `List`. It's declared like this:

```
public inline fun <T> mutableListOf(): MutableList<T> = ArrayList()
```

That likely doesn't all make sense yet, but here's the gist of it: this is a method on `Collection` that takes in a list of objects and returns a `MutableList`. `MutableList` is a subclass of `MutableCollection`, which in turn inherits from `Collection`.

It's helpful to get your head around how all these things relate to each other, so Figure 5.1 shows a partial inheritance tree starting with `Collection`.

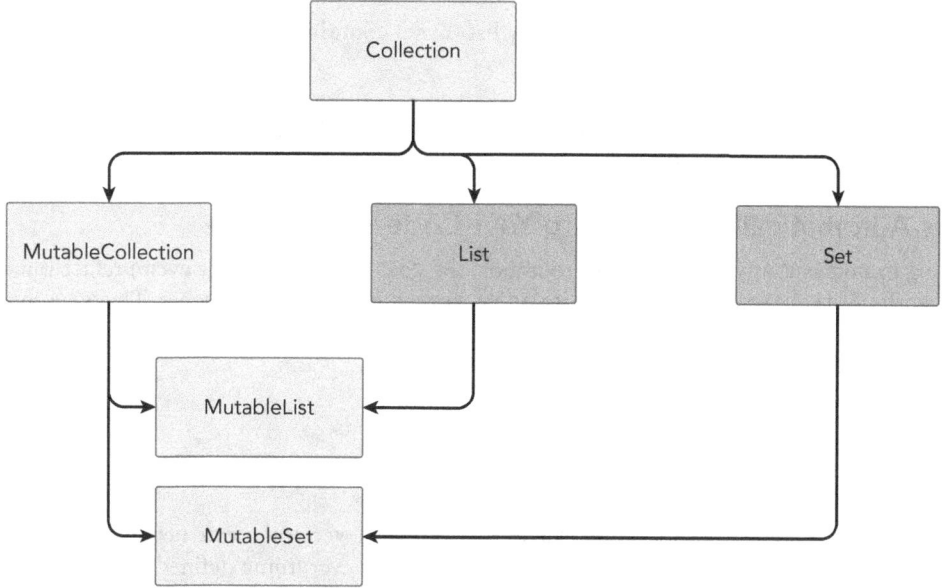

FIGURE 5.1: `Collection` and several related classes

This isn't all the classes related here (there's `Iterable`, `MutableIterable`, and `Map` and `MutableMap`), but for now, these are the ones on which to focus. Notice that a list that can be changed is in fact an entirely different class than one that cannot (`MutableList` versus `List`). This is a great example of the principle discussed in the previous chapter: the base class is more general, and the subclass is more specific. That's what you're seeing here in practice.

> **WARNING** *There's quite a bit in Figure 5.1 that may strike you as odd. Most notably, there are two arrows going to* `MutableList` *and* `MutableSet`*: one to each from* `MutableCollection` *and one to each from* `List` *or* `Set`*. How does that work? Can a single class have two superclasses? That's a good question, and one you'll dig into in the next chapter.*

When you call `mutableListOf`, you're calling a method on `Collection`. It creates a `MutableList` (and specifically, an `ArrayList`) and returns that to your code. That method is basically a *factory* for lists (that's where the name "factory pattern" comes from).

As you'd expect, there are other factory methods on `Collection`:

➤ `listOf()`: Creates a new read-only list (`List`) containing what you pass into the method

➤ `emptyList()`: Creates a new empty list

➤ `mutableListOf()`: Creates a new mutable list (`MutableList`) containing what you pass into the method

➤ `listOfNotNull()`: Creates a new read-only list (`List`) containing all non-null objects passed into this method

You'll use many of these as you work through this chapter. In each case, you use the factory method to create an instance of the object you need.

Collection Is Automatically Available to Your Code

Another thing to note is that you didn't have to import the `Collection` class, or even preface `muta-bleListOf` with `Collection`, like you might if you've worked with other languages. That's because of three things:

➤ `Collection` is in the `kotlin.collections` package.

➤ The factory methods in `Collection` are all public methods.

➤ `Collections` is really just a collection of classes and methods.

Kotlin automatically imports a number of packages for you whenever you create new Kotlin code. One of those packages is `kotlin.collections`. That means that everything defined in that package—including `Collection`—is immediately available for use within your program without an import statement.

Second, the various factory methods are all `public`, like this:

```
public fun <T> emptyList(): List<T> = EmptyList
```

That means you can use them immediately, as they are visible to your code.

And third, there is a difference between a `Collection`, which is defined by Kotlin, and the *Collections* file, the actual source code file. The source code defines a number of methods—like these factory methods—that are *not* in a class. They're simply public methods available to your program. That's why you don't have to write code like this:

```
val rockBands = Collection.mutableListOf("The Black Crowes", "Led Zeppelin",
                                         "The Beatles")
```

> **NOTE** *If this is a little confusing, that's normal. The* `Collections` *code and package does some things that are more about Kotlin internals than you need to worry about right now. Just realize that you can access methods defined in the* `Collections` *file because that file is in an automatically imported package, the methods are public, and they're not defined as part of an actual class. That's all you need to know for now.*

Mutating a Mutable List

Back to your list of bands:

```
val rockBands = mutableListOf("The Black Crowes", "Led Zeppelin",
        "The Beatles")
```

One thing you should do right away is figure out exactly what you have. This isn't specific to collections in Kotlin, but you can always use this notation to get a variable's type in Kotlin:

```
println("rockBands is a ${rockBands::class.simpleName}")
```

For rockBands, this will report as an ArrayList, something mentioned earlier. ArrayList is simply an implementation of a MutableList. And because your list is mutable, you can use the add() method to add to it:

```
rockBands.add("The Rolling Stones")
println(rockBands)
```

> **NOTE** *For the most part, output won't be shown for these simple operations. The* println() *statement is more to remind you that you should be checking to see what these operations do yourself. Output will still be shown if it's particularly surprising or notable, though.*

You can also remove items by their index in the list:

```
rockBands.removeAt(2)
println(rockBands)
```

Lists in Kotlin, like most languages, are zero-based. That means that if you have this list:

```
[The Black Crowes, Led Zeppelin, The Beatles, The Rolling Stones]
```

the first item is at index 0. So removeAt(2) removes the third item in the list, because the first is index 0, the second is index 1, and the third is index 2.

Getting Properties from a Mutable List

You're also going to want to get used to seeing how big a list is and directly pulling values from it. Listing 5.2 shows a number of common operations in action.

LISTING 5.2: Working with mutable lists through basic methods

```
fun main() {
    val rockBands = mutableListOf("The Black Crowes", "Led Zeppelin",
                        "The Beatles")
    println(rockBands)
    println("rockBands is a ${rockBands::class.simpleName}")

    rockBands.add("The Rolling Stones")
    println(rockBands)
```

continues

LISTING 5.2 *(continued)*

```
println("There are ${rockBands.size} bands in the list.")
rockBands.removeAt(2)
println(rockBands)
println("Now there are ${rockBands.size} bands in the list.")

// Remember that lists are zero-based!
println("The first band is ${rockBands.get(0)}")

// Use array notation
println("The second band is ${rockBands[1]}")

// Add at a specific index
rockBands.add(1, "The Eagles")
println("The second band is ${rockBands[1]}")
println(rockBands)

}
```

Nothing here should be very surprising, especially if you've worked with lists in other programming languages.

> **NOTE** *All of these properties and methods are available on any* MutableList *and aren't specific to* ArrayList*. Also, for the record, an* Array *in Kotlin is a statically sized array; it can't grow or shrink after creation. An* ArrayList*, then, is really just a resizable array.*

LISTS (AND COLLECTIONS) CAN BE TYPED

So far, you've just thrown a bunch of strings into a list, and Kotlin was happy to comply. But you can throw different types of things into a list as well:

```
val mixedThings = mutableListOf("Electric Guitar", 42)
println(mixedThings)
```

This doesn't make much sense, because that list is now not very useful. You can't iterate over it (something you're going to do very soon and do a *lot*) and treat each object the same, because they're not the same. (In this case, you can't do things to a number that you can do to a string, and vice versa.) You're also going to have to likely check the type of what you get every time you pull something out of the list, like this:

```
val mixedThings = mutableListOf("Electric Guitar", 42)
println(mixedThings)

val item = mixedThings[0]
if (item is String) {
    println("Value: ${item}")
```

```
    } else if (item is Int) {
        println(3 + item)
    }
```

That's not a great idea, and you can do much better.

Give Your Lists Types

A better idea is to tell Kotlin what types you'll allow in your list when you create the list. Listing 5.3 is a revised version of `PersonTest` (from the previous chapters) that uses a mutable list, and tells Kotlin that it can only have `Person` instances within it.

LISTING 5.3: Creating a list with a type

```
import org.wiley.kotlin.person.Person
import org.wiley.kotlin.person.Parent

fun main() {
    // Create a new person
    val brian = Person("Brian", "Truesby", 68.2, 33)
    printPersonPartnerStatus(brian)

    // Create another person
    val rose = Person("Rose", "Bushnell", brian)
    printPersonPartnerStatus(rose)

    // Change Rose's last name
    rose.lastName = "Bushnell-Truesby"
    println(rose)

    val mom = Parent("Cheryl", "Truesby", _child=brian)
    mom.child = rose

    val somePeople = mutableListOf<Person>(brian, rose)
    println(somePeople)

    // Subclasses of Person are still a Person!
    somePeople.add(mom)
    println(somePeople)

    // Create a person and add it in one step
    somePeople.add(Person("Brett", "McLaughlin"))
    println(somePeople)

    // Only Person instances can be added
    somePeople.add(32)
}

fun printPersonPartnerStatus(person: Person) {
    if (person.hasPartner()) {
        println("$person has a partner: ${person.partner}.")
    } else {
        println("$person does not have a partner.")
    }
}
```

This will not compile, because of that last line in `main` trying to add an `Int` (32) to the list. Remove that line and things will work, and you'll get output like this:

```
[Brian Truesby, Rose Bushnell-Truesby]
[Brian Truesby, Rose Bushnell-Truesby, Cheryl Truesby]
[Brian Truesby, Rose Bushnell-Truesby, Cheryl Truesby, Brett McLaughlin]
```

Notice that any object that's a `Person` instance works—including subclasses of `Person`. And of course, this type checking applies to all of the methods on `List` (or `Set`, when we get to sets) that add objects.

You can use several more methods to operate on lists, but when you introduce typing, there's one thing you're going to do *a lot*: iterate over a list.

Iterate over Your Lists

When you have a list, you generally type it for one reason: you want to work through that list (often one item at a time) and treat each item as a certain thing—the type. Then you can do something based on knowing that type. To demonstrate this, you'll need a little more code.

Listing 5.4 introduces a very simple new class, `Band`. It's horribly limited (whoever heard of a band that has no place for keys? And where's the lead guitar player?), but it is a useful example.

LISTING 5.4: Very simple new class to represent a Band

```
package org.wiley.kotlin.music

import org.wiley.kotlin.person.*

class Band(_name: String, _singer: Person, _guitarist: Person, _drummer: Person, _
bassist: Person) {

    var name: String = _name
    var singer: Person = _singer
    var guitarist: Person = _guitarist
    var drummer: Person = _drummer
    var bassist: Person = _bassist
}
```

This class introduces a new package, `org.wiley.kotlin.person`, and also uses the `Person` class from previous chapters.

Now go back to your test class and update that code to use this class, as shown in in Listing 5.5.

LISTING 5.5: Reworking CollectionTest to use the new Band class

```
import org.wiley.kotlin.music.*
import org.wiley.kotlin.person.*

fun main() {
    val rockBands = mutableListOf<Band>(
```

```
Band(
    "The Beatles",
    _singer = Person("John", "Lennon"),
    _guitarist = Person("George", "Harrison"),
    _bassist = Person("Paul", "McCartney"),
    _drummer = Person("Ringo", "Starr")
),
Band(
    "The Rolling Stones",
    _singer = Person("Mick", "Jagger"),
    _guitarist = Person("Keith", "Richards"),
    _bassist = Person("Ronnie", "Wood"),
    _drummer = Person("Charlie", "Watts")
),
Band(
    "The Black Crowes",
    _singer = Person("Chris", "Robinson"),
    _guitarist = Person("Rich", "Robinson"),
    _drummer = Person("Steve", "Gorman"),
    _bassist = Person("Johnny", "Colt")
)
)

}
```

> **NOTE** *The order of inputs to the* Band *constructor is adjusted throughout the code by using named parameters. There's no real method behind the ordering; it was simply a matter of the order that band members were put into the object instance's constructor (and honestly that was a factor of the order in which I remembered their names!).*

This might look a bit different but is really nothing new at this stage. rockBands is defined as a list of Band objects, and then each Band is created inline, with the members created as Person instances within the Band constructor.

Now you can do all the same things with rockBands you did when it just held strings:

```
println(rockBands)
println("rockBands is a ${rockBands::class.simpleName}")
println("There are ${rockBands.size} bands in the list.")
rockBands.add(
    Band(
        "The Black Keys",
        _singer = Person("Dan", "Auerbach"),
        _guitarist = Person("Steve", "Marion"),
        _drummer = Person("Patrick", "Carney"),
        _bassist = Person("Richard", "Swift")
    )
)
```

```
println("Now there are ${rockBands.size} bands in the list.")
```

```
println("The second band is ${rockBands[1]}")
```

The output here is a bit underwhelming, though:

```
[org.wiley.kotlin.music.Band@2812cbfa, org.wiley.kotlin.music.Band@2acf57e3,
org.wiley.kotlin.music.Band@506e6d5e]
rockBands is a ArrayList
There are 3 bands in the list.
Now there are 4 bands in the list.
The second band is org.wiley.kotlin.music.Band@2acf57e3
```

The problem, even for these simple test purposes, is that a Band doesn't have a useful toString() method. You can fix that with a very simplistic implementation in Band:

```
override fun toString(): String {
    return name
}
```

Rerun your test and you'll get some useful output:

```
[The Beatles, The Rolling Stones, The Black Crowes]
rockBands is a ArrayList
There are 3 bands in the list.
Now there are 4 bands in the list.
The second band is The Rolling Stones
```

Now you have something to iterate over: a list of bands. The easiest way to work through this list is to get an iterator from your list itself:

```
val bandsIterator = rockBands.iterator()
```

This really makes sense; a list is the best source for getting the right sort of iterator. What you get back is an object (a subclass of Iterator) that knows how to move through your list—in this case, rockBands.

You can use a while loop, something new to this book but also something you've likely seen before in other languages, to loop using this iterator:

```
val bandsIterator = rockBands.iterator()
while (bandsIterator.hasNext()) {
    var band = bandsIterator.next()
    // do something with the next band
}
```

As long as hasNext() is true, there's another band in the list. You can get that band with next(), which—because of Kotlin's strong typing—you know will be a Band instance. Then you can print (or do anything else with) that Band's properties as needed:

```
val bandsIterator = rockBands.iterator()
while (bandsIterator.hasNext()) {
    var band = bandsIterator.next()
    println("The lead singer for ${band.name} is ${band.singer}")
}
```

All of this works because of this little bit of code from earlier:

```
val rockBands = mutableListOf<Band>
```

That typing is the key to the iterator and the use of each object in the list as a `Band` working.

Kotlin Tries to Figure Out What You Mean

Two different shortcuts are available to you here, and both are worth noting because they can make your code simpler and cleaner. The first is that Kotlin will infer what type of list you have if you feed the list factory construction method objects of the same type.

So go back to your `main()` code and remove the `<Band>` part of this line:

```
val rockBands = mutableListOf(
```

Now compile, and everything still works perfectly as it should.

Additionally, there's a shortcut to getting an iterator and then using a `while` loop. Instead, you can use a `for` loop and the `in` keyword, like this:

```
for (band in rockBands) {
    println("The lead singer for ${band.name} is ${band.singer}")
}
```

This saves a few lines of code. This will loop through every item in the `rockBands` list. Each item is assigned to a variable called `band` that is available within the loop body (between the { and }).

So should you use both of these shortcuts? Well, maybe not. The `for` loop is a great idea, and your code is very readable. But look back at that factory method for creating a list:

```
val rockBands = mutableListOf(
```

There's nothing wrong here, but this actually tells you (and anyone else reading your code) *less* about your list. You don't know that it's supposed to only contain `Band` objects. So in many cases, adding the type actually makes this code clearer, and it only requires a few extra keystrokes; that's a pretty good tradeoff.

```
val rockBands = mutableListOf<Band>(
```

> **NOTE** *There's another really good reason to type a list like this related to inheritance and subclasses. You'll see that soon, but keep in mind that a list will infer the type of the list by the common class of the items added to it—not a common base or superclass. If that doesn't trigger any alarm bells, don't worry about it for now. You'll see soon how specifying a type can save you some trouble down the line.*

LISTS ARE ORDERED AND CAN REPEAT

Probably the most important thing to know about a list, and why you'd use it over a set or map (both of which you'll see soon) is that a list is ordered, and a list can contain the same item twice. A set is not ordered and cannot store duplicate items. A map is actually quite a bit different than either and uses key-value pairs.

You'll quickly find it's easy to determine which type you need, and lists are often most common. If you want to have something that behaves like an array—items are in an order, items can be accessed by index, and you don't care about duplicates—then go with a list.

Order Gives You Ordered Access

You've already seen two ways to access an item in a list by index:

```
println("The second band is ${rockBands[1]}")
println("The third band is ${rockBands.get(2)}")
```

You can also grab the first or last item in a list:

```
println("The first band is ${rockBands.first()}")
println("The second band is ${rockBands[1]}")
println("The last band is ${rockBands.last()}")
```

None of these methods will apply to non-ordered collections. You can also shuffle list items:

```
println("The first band is ${rockBands.first()}")
println("The second band is ${rockBands[1]}")
println("The last band is ${rockBands.last()}")

rockBands.shuffle()

println("The first band (post shuffle) is ${rockBands.first()}")
println("The second band (post shuffle) is ${rockBands[1]}")
println("The last band (post shuffle) is ${rockBands.last()}")
```

The output here reflects the shuffling:

```
The first band is The Beatles
The second band is The Rolling Stones
The last band is The Black Keys
The first band (post shuffle) is The Black Keys
The second band (post shuffle) is The Beatles
The last band (post shuffle) is The Rolling Stones
```

Lists Can Contain Duplicate Items

You can also add the same item to a list multiple times. Here's a trivial example:

```
var kinks = Band("The Kinks",
                 _singer = Person("Ray", "Davies"),
                 _guitarist = Person("Dave", "Davies"),
                 _drummer = Person("Mick", "Avory"),
                 _bassist = Person("Pete", "Quaife")
)

rockBands.add(kinks)
rockBands.add(kinks)
rockBands.add(1, kinks)
```

Now you'll have three different elements in the list with the Kinks:

```
The lead singer for The Beatles is John Lennon
The lead singer for The Kinks is Ray Davies
The lead singer for The Black Keys is Dan Auerbach
The lead singer for The Rolling Stones is Mick Jagger
The lead singer for The Black Crowes is Chris Robinson
The lead singer for The Kinks is Ray Davies
The lead singer for The Kinks is Ray Davies
```

This isn't a problem as long as you know it's possible, and don't assume your list has only unique elements. If you *do* want to only deal with unique elements, you can use the distinct() method:

```
for (band in rockBands.distinct()) {
    println("The lead singer for ${band.name} is ${band.singer}")
}
```

Now the output only has each band once:

```
The lead singer for The Beatles is John Lennon
The lead singer for The Kinks is Ray Davies
The lead singer for The Rolling Stones is Mick Jagger
The lead singer for The Black Crowes is Chris Robinson
The lead singer for The Black Keys is Dan Auerbach
```

This method is what is called nondestructive, though. It does *not* actually change the list it's called on. Instead, it returns a new list. So you could store that new list:

```
var singularBands = rockBands.distinct()
```

In this case, you'd have rockBands, which has (potentially) duplicates, and singularBands, which will absolutely *not* have any duplicates.

> **NOTE** *You can still do a lot more things with lists. In fact, you could probably read another 20 or 30 pages before you'd hit the extent of it all: filtering, sorting, selecting uniqueness based on specific attributes, and much more. This is plenty for you to make good practical use of lists, though, and as more needs arise, you can easily add to your lists and collections toolkit.*

SETS: UNORDERED BUT UNIQUE

You already know a lot of what you need to know about sets. Like lists, they inherit from Collection, and like lists, they can be mutable or read-only. They also are commonly iterated over, something you just learned how to do.

The big differences are that a set is unordered (while a list is ordered) and a set has one and only one of every item in it (while a list can have duplicates).

In Sets, Ordering Is Not Guaranteed

Yes, you've now heard multiple times that sets don't preserve order. But they *might*. In other words, sets don't say "your ordering will be random." Sets simply say, "no promises. We can do whatever we want with ordering of elements."

For example, Kotlin will typically default to supplying you a `LinkedHashSet`:

```
val bandSet = setOf<Band>(rockBands[0], rockBands[1],
                          rockBands[2], rockBands[3])
println("${bandSet::class.simpleName}")

println(rockBands)
println(bandSet)
```

You'll see that `bandSet` here (which contains the first four elements from your `rockBands` list) is a `LinkedHashSet`:

```
LinkedHashSet
[The Beatles, The Kinks, The Rolling Stones, The Black Crowes, The Black Keys, The
Kinks, The Kinks, Bad Company]
[The Beatles, The Kinks, The Rolling Stones, The Black Crowes]
```

> **NOTE** *Note that you may see these items in a different order. Don't worry about that; it's not an issue that indicates that something is wrong.*

`LinkedHashSet` happens to use the ordering in which elements were added. So if you compare its first four elements to the first four of `rockBands`, you'll see that they match.

But that's not the same as a guarantee of ordering! `rockBands` will always report ordering; `bandSet` can change that order anytime it wants without violating any of Kotlin's rules.

But now use a `HashSet` (with `hashSetOf()`) and see the difference:

```
val bandSet = hashSetOf<Band>(rockBands[0], rockBands[1],
                              rockBands[2], rockBands[3])
println("${bandSet::class.simpleName}")

println(rockBands)
println(bandSet)
```

The output here will not show matching first elements:

```
HashSet
[The Beatles, The Kinks, The Rolling Stones, The Black Crowes, The Black Keys, The
Kinks, The Kinks, Bad Company]
[The Beatles, The Rolling Stones, The Black Crowes, The Kinks]
```

`HashSet` does not preserve order.

When Does Order Matter?

Given that sets don't order items and lists do, why would you use one over the other? Well, most notably, sets that don't order items are typically faster when seeing if an item is in that set.

> **WARNING** *Beware that there is a difference between a set that doesn't have to maintain order and a set that actually doesn't maintain order. The default set,* LinkedHashSet, *doesn't have to maintain order but still does. That means that it's not going to be significantly faster in these respects than a list. If you're looking to use a set primarily for speed purposes, use* HashSet *via* hashSetOf() *to be sure you get an unordered set.*

This mostly has to do with the use of hashes in sets (thus, HashSet). But this means you've made a commitment that order doesn't matter, and that's not always obvious. It's going to depend both on what you are modeling and how you are using the set.

For example, take the list of rock bands. If you are just collecting a list of bands, you really don't need ordering, so a set is a great choice. But what if you are doing more than just collecting bands. What if:

➤ You are collecting the bands ordered by debut album release?

➤ You are using the list as a set list of bands and the order is the order in which they play?

➤ You are alphabetizing the bands?

In these cases, you *might* want to use a list to preserve order.

But why might? Aren't these all situations where order matters? Yes, but that may not require you to preserve that ordering *within the list structure*.

Sort Lists (and Sets) on the Fly

Lists and sets provide a method called sortedBy() that allows you to sort a list or set without worrying about the actual ordering of that list. Here's an example that sorts the bands list by the name of the band:

```
println(rockBands.sortedBy { band: Band -> band.name })
```

So you call sortedBy() and then give it a block of code. You're identifying a variable to represent the item at each list index (which in this case we know is a Band, thanks to Kotlin strong typing), and then to indicate a property to sort on. In this case, the band's name is used.

The result is a sorted output:

```
[Bad Company, The Beatles, The Black Crowes, The Black Keys,
The Kinks, The Kinks, The Kinks, The Rolling Stones]
```

You could save this new list to a variable, or just use it directly as is the case here.

So then why ever use a list at all? Well, you might want to keep up with the ordering based on when something was added to the list. Or you might want to preserve ordering because you don't have an attribute to sort by in the objects within the actual list. In these cases, you want the ordering, and you want to count on the ordering.

Sets: No Duplicates, No Matter What

Another reason you might want to use a set over a list is that sets don't allow duplicates. Go ahead and try this and see what happens:

```
val bandSet = hashSetOf<Band>(rockBands[0], rockBands[1],
                              rockBands[2], rockBands[3])
println("${bandSet::class.simpleName}")

// Add a band that's already been added
bandSet.add(rockBands[0])
```

You will *not* get an error from Kotlin here; it will happily run this code. But the output reveals that there's still only one of each item in the resulting set:

```
[The Beatles, The Rolling Stones, The Black Crowes, The Kinks]
```

Sets "Swallow Up" Duplicates

You'll notice here something important: it's not an error to attempt to add a duplicate to a set; it's just ignored. You won't see an error; you just also won't see the addition.

This is important, because if you were expecting an issue or even a notification, you won't get it. You can, however, use the contains() method on a set to see if it contains an element. You can also use this as a check before adding a new item:

```
// Add a band if it's not already in the set
if (bandSet.contains(rockBands[0])) {
    println("bandSet already contains ${rockBands[0]}")
} else {
    bandSet.add(rockBands[0])
}
```

Sets Use equals(x) to Determine Existing Membership

All of this talk of duplicates raises an important question: how does a set determine if an item is a duplicate? The answer is actually easy, and you may have already thought about it: equals(x) is used. So if a new item equals any existing item—based on return values from equals(x)—then that item isn't added to the set.

To see this fail when you might think it would *not* fail, look at this chunk of code:

```
var duplicateBand = rockBands[0]

// Add a band if it's not already in the set
if (bandSet.contains(duplicateBand)) {
    println("bandSet already contains ${duplicateBand}")
} else {
    bandSet.add(duplicateBand)
}
```

```
var bandTwo = Band(
    rockBands[0].name,
    _singer = Person(rockBands[0].singer.firstName,
                     rockBands[0].singer.lastName),
    _guitarist = Person(rockBands[0].guitarist.firstName,
                        rockBands[0].guitarist.lastName),
    _bassist = Person(rockBands[0].bassist.firstName,
                      rockBands[0].bassist.lastName),
    _drummer = Person(rockBands[0].drummer.firstName,
                      rockBands[0].drummer.lastName)
)

// Add a band if it's not already in the set
if (bandSet.contains(bandTwo)) {
    println("bandSet already contains ${bandTwo}")
} else {
    bandSet.add(bandTwo)
}
```

Take some time to make sure you see what's happening here. First, a duplicate is added by taking an element already in the set (from earlier in the code) and adding it to the set. As you'd expect, that element won't get added (it turns out to be the Beatles in my case, by the way).

> **NOTE** *Like the order of bands earlier, you may see a different band at issue here. As before, this isn't a problem.*

But then, another Band instance is created, and given the name and each individual instance data for each member of the band:

```
var bandTwo = Band(
    rockBands[0].name,
    _singer = Person(rockBands[0].singer.firstName,
                     rockBands[0].singer.lastName),
    _guitarist = Person(rockBands[0].guitarist.firstName,
                        rockBands[0].guitarist.lastName),
    _bassist = Person(rockBands[0].bassist.firstName,
                      rockBands[0].bassist.lastName),
    _drummer = Person(rockBands[0].drummer.firstName,
                      rockBands[0].drummer.lastName)
)
```

In theory, rockBands[0] should be the equivalent of bandTwo, so when bandTwo is added to band-Set, it should get ignored. But check the output; that's not what happened:

```
[The Beatles, The Beatles, The Black Crowes, The Kinks, The Rolling Stones]
```

So what gives? Well, the answer lies in equals(x). Even though the data in these two objects is identical, they are not the same object. You should recall from Chapter 3 that without a custom implementation of equals(x), object equality is typically based on memory location of an object—so two objects with the same data are *not* equal.

You can see this is true if you make some simple fixes to `equals(x)`. Listing 5.6 shows an update to `Band` that addresses this (and includes a matching change to `hashCode()`).

LISTING 5.6: Updated equals(x) and hashCode() to improve Band

```kotlin
package org.wiley.kotlin.music

import org.wiley.kotlin.person.*

class Band(_name: String, _singer: Person, _guitarist: Person, _drummer: Person,
_bassist: Person) {

    var name: String = _name
    var singer: Person = _singer
    var guitarist: Person = _singer
    var drummer: Person = _drummer
    var bassist: Person = _bassist

    override fun toString(): String {
        return name
    }

    override fun hashCode(): Int {
        return name.hashCode()
    }

    override fun equals(other: Any?): Boolean {
        if (other is Band) {
            return (name == other.name)
        } else {
            return false
        }
    }
}
```

> **NOTE** *Like the* `equals(x)` *method in* `Person` *from earlier chapters, this is a bit simplistic. (It's actually quite a bit more simplistic than* `Person`.*) Still, it illustrates the point in question.*

Now, rerun your test code, and you'll see only a single instance of the Beatles:

```
[The Beatles, The Black Crowes, The Kinks, The Rolling Stones]
```

With an `equals(x)` method that reports that the two instances are equal based on band name, the set now only accepts the first addition.

Using a Set? Check equals(x)

The final thing that this illustrates is that if you're filling a set with any custom objects, you need to be sure that equals(x) for those objects does what you think it does. There's nothing harder to track down than a set that you chose to eliminate duplicates not actually appearing to eliminate duplicates.

When in doubt, look at the source code for custom objects. Or at a minimum, you can compare a couple of instances that you would assume as equal in code using equals(x) and see what happens. This will at least let you prepare accordingly.

Iterators Aren't (Always) Mutable

One thing you haven't seen is that the Iterator object you get from a list or a set allows you to change the collection that the Iterator is operating upon:

```
var iterator = rockBands.iterator()
while (iterator.hasNext()) {
    var band = iterator.next()
    if (band.name.contains("black", ignoreCase = true)) {
        iterator.remove()
    }
}
```

If you compile and run this code, and then print rockBands after its execution, you will see that the Black Keys and the Black Crowes have both been removed.

The reason this works is twofold: a collection's iterator knows about the underlying collection, so can remove or add things as needed; and a mutable collection gets a mutable iterator. In fact, MutableIterator inherits from Iterator. A collection that is *not* mutable does *not* provide a mutable iterator, for obvious reasons: the underlying collection can't change.

MAPS: WHEN A SINGLE VALUE ISN'T ENOUGH

So far, you've seen lists and sets. They're pretty similar: they store a collection of items. It's only in how they store those items, and whether order and duplication matters, that set these apart. But there's another collection type that is quite different: Map.

A Map stores *key-value pairs*. So for every item in the map, there's a key (a unique identifier) and a value (what that identifier points to, or is associated with). Generally, the key is *not* just some property of the value, as that's somewhat redundant. So you wouldn't have a map, for example, where the key was a person's first name and the value was a Person object with that first name.

You *will* often see a map with a numeric ID of some sort and a value associated with that ID, often an object.

Maps Are Created by Factories

Like a set or list, you create a map with a factory method. Unlike those two collections, though, you have to provide both key and value, and the notation looks a bit different:

```
var statusCodes = mapOf(200 to "OK",
                        202 to "Accepted",
                        402 to "Payment Required",
                        404 to "Not Found",
                        500 to "Internal Server Error")
```

> **NOTE** *This particular map is modeling the various HTTP status codes and the associated description. It's also wholly incomplete. There are many, many more status codes.*

While this might appear odd, think about the phrase "maps to" for that to keyword. So here, 202 maps to "Accepted" and 402 maps to "Payment Required."

You can print out the keys and values with those properties on the map:

```
var statusCodes = mapOf(200 to "OK",
                        202 to "Accepted",
                        402 to "Payment Required",
                        404 to "Not Found",
                        500 to "Internal Server Error")
println("Keys: ${statusCodes.keys}")
println("Values: ${statusCodes.values}")
```

Now when it comes to actually getting an individual value, you use the same sorts of methods as with lists and sets:

```
println("202 has an associated value of '${statusCodes.get(202)}'")
```

Use Keys to Find Values

Under the hood, your new Map is actually either a Set of key-value pairs or a List of key-value pairs. That means that order may or may not be preserved in your map. In other words, you should *not* be writing code that depends on the ordering of keys in a map.

Instead, you should be using keys to find values, or looking up keys or values, as shown here:

```
if (statusCodes.contains(404)) {
    println("Message: ${statusCodes.get(404)}")
} else {
    println("No message for 404.")
}
```

Of course, you need a key to get a value. If you supply a key that doesn't map to a value, you won't get anything. In fact, you'll get a null value:

```
println(statusCodes.get(400))
```

This returns `null`, meaning you'll likely need code that either checks for a key before calling `get()` using `contains()`, or handles the possibility of `null` return values.

How Do You Want Your Value?

If you want to avoid that—and in many cases, it's not worth all the extra code—you can use `getOr-Default()`. This variation of `get()` allows you to specify a return value to use if `get()` would normally return `null`:

```
println(statusCodes.getOrDefault(400, "Default Value"))
```

Here, assuming 400 isn't a key in the set, you'll simply get back "Default Value."

Another option is to use `getValue()`, which works just like `get()` but actually throws an exception if the key specified isn't in the map:

```
println(statusCodes.getValue(400))
```

Run this code and you'll get a runtime exception:

```
Exception in thread "main" java.util.NoSuchElementException:
        Key 400 is missing in the map.
    at kotlin.collections.MapsKt__MapWithDefaultKt.getOrImplicitDefaultNullable(
    MapWithDefault.kt:24)
    at kotlin.collections.MapsKt__MapsKt.getValue(Maps.kt:336)
    at CollectionTestKt.main(CollectionTest.kt:140)
    at CollectionTestKt.main(CollectionTest.kt)
```

That's pretty heavy-handed, so you may not want to use that unless you really do want to stop everything in the event of a non-existent key.

FILTER A COLLECTION BY . . . ANYTHING

So far, you've seen how to take a collection—`List`, `Set`, or `Map`—and iterate over it, select items from it, grab a certain value at a certain index or position, and generally work with these collections freely. However, one thing that hasn't been discussed is getting a collection of elements from within the collection based on certain criteria.

Suppose you have the `rockBands` list and you want all the bands from the list with the word "Black" as part of their name. One approach would be to use an `Iterator` and simply create a new list and fill it as you go through the list. You've already written code to do something like this:

```
var bandsInBlack = mutableListOf<Band>()
for (band in rockBands) {
    println(band.name)
    if (band.name.contains("Black", ignoreCase = true)) {
        bandsInBlack.add(band)
    }
}
println(bandsInBlack)
```

> **WARNING** *If you've been following along and just adding the sample code to your collection test, you may not get the expected results here. There's an earlier snippet of code that removes all bands in* rockBands *that have a name containing "Black," so you'll need to either remove or comment out that code for this to do what it should.*

There's certainly nothing wrong with this code, but this is such a common operation that Kotlin gives you some shortcuts to get the same result.

Filter Based on a Certain Criterion

The `filter()` method is your best friend when you want to select all the elements in a collection based on a criteria. So take the same task: you want all the bands from `rockBands` that have a name containing the word "Black." This is ideal for `filter`:

```
println(rockBands.filter { it.name.contains("Black", ignoreCase = true) })
```

This code probably is pretty self-explanatory, but walk through it to make sure you know exactly what's going on.

First, `filter` is a bit different in that it takes a code block surrounded by { and } instead of parentheses with arguments. Within that code block, you have access to an object instance named `it`. This `it` is somewhat like the `field` and `value` in a customer mutator or accessor; they're stand-ins for real object instances or field values with intentionally generic names.

Here, `it` represents each object instance in the collection; in this case, a list. So you can think of `it` as standing in for a `Band` instance, of which the list has some number. So for each `Band` instance, evaluate the `name` property to see if it contains "Black" (while ignoring case). If the entire expression evaluates to true, then the object instance is included as part of the returned list of values; if it's false, then the object instance isn't included.

You can make the evaluation as complex as you need. Here's another one that returns bands where either the bassist or singer has a first name of "Paul":

```
println(rockBands.filter {
    (it.singer.firstName.equals("Paul", ignoreCase = true)) ||
    (it.bassist.firstName.equals("Paul", ignoreCase = true))
})
```

There's nothing particularly tricky here. It's just a matter of constructing the right Boolean expression.

This expression looks a bit different for a `Map`, but works in essentially the same way. You'll just have access to both the key and value for each item in the map, rather than just the list or set value:

```
println(statusCodes.filter { (key, value) -> key in 400..499 })
```

Here, the `key` is used to only return values between 400 and 499. You can also use the `value`:

```
println(statusCodes.filter { (key, value) -> key in 400..499 &&
                             value.startsWith("Payment") })
```

Again, as long as you can express the criteria as a Boolean expression, you can use `filter()`.

> **NOTE** `filter()` *is still written with parentheses as a shorthand to indicate that it's a method. But it actually takes what is called a* predicate, *which is the expression you pass to it to be evaluated.*

Filter Has a Number of Useful Variations

You can do more with `filter()` once you understand its basic usage and how to write predicates. One useful one is `filterNot()`, which returns the inverse of what `filter()` returns. So where `filter()` returns elements that evaluate to true in the predicate, `filterNot()` returns elements that evaluate to false.

Here is a simple filter that returns all bands that do *not* have "Black" in their name:

```
println(rockBands.filterNot { it.name.contains("Black", ignoreCase = true) })
```

> **NOTE** *It's generally safe to say that* `filterNot()` *returns the things that* `filter()` *would not, and vice versa, but in some special cases both might evaluate to not include the element, most notably if there's a null value involved.*

Another interesting variation on `filter()` is `filterIsInstance()`. This variation allows you to select all the elements from a collection that are of a certain type.

This might seem odd when you consider that most collections are strongly typed; what would this help with? But it's actually particularly useful with strongly typed collections of custom objects that have subclasses. For example, suppose that you developed a custom `RockBand` class, `JazzBand` class, and `CountryBand` class, all of which inherited from `Band`. You could use `filterIsInstance()` to take a list typed to `Band` and return all the members of the list that are specifically `RockBand`:

```
var justRock = bands.filterIsInstance<RockBand>()
```

Now you have a new typed list of just the subtype you want.

COLLECTIONS: FOR PRIMITIVE AND CUSTOM TYPES

Most of the examples in this chapter have used a combination of Kotlin types. *Primitive* types, like `Int`, `Float`, `String`, and `Boolean`, are all easy to work with in collections. You've also worked with collections of custom objects (`Band` most notably, as well as `Person`). They're no more complex to work with, but you will need to ensure that you have useful `equals(x)` and `hashCode()` methods and you understand the types so you can work with them appropriately.

> **NOTE** *A primitive type is a programming term for a basic type. Usually, that means a type that can't be further broken down, or that doesn't have individual parts. Custom types—like custom objects—are not primitive. They're less basic and they often have their own properties.*

Much of the work with collections really is about understanding how to use them at a functional level, not a mechanical level. In other words, you learned how to iterate over, add to, and remove from a collection in some 20 pages; it will take you a *lot* of programming with them to get a great sense of when they are best used, and when they're not, and when to use a map or set or list.

Add a Collection to Person

With that in mind, it's time to revisit `Person`, and specifically, the subclass `Parent` from Chapter 4. If you recall, `Person` looked like Listing 5.7 and `Parent` like Listing 5.8.

LISTING 5.7: The Person class from Chapter 4 (and before)

```kotlin
package org.wiley.kotlin.person

open class Person(_firstName: String, _lastName: String,
                  _height: Double = 0.0, _age: Int = 0) {

    var firstName: String = _firstName
    var lastName: String = _lastName
    var height: Double = _height
    var age: Int = _age

    var partner: Person? = null

    constructor(_firstName: String, _lastName: String,
                _partner: Person) :
                this(_firstName, _lastName) {
        partner = _partner
    }

    fun fullName(): String {
        return "$firstName $lastName"
    }

    fun hasPartner(): Boolean {
        return (partner != null)
    }

    override fun toString(): String {
        return fullName()
    }
}
```

```kotlin
    override fun hashCode(): Int {
        return (firstName.hashCode() * 28) + (lastName.hashCode() * 31)
    }

    override fun equals(other: Any?): Boolean {
        if (other is Person) {
            return (firstName == other.firstName) &&
                    (lastName == other.lastName)
        } else {
            return false
        }
    }
}
```

LISTING 5.8: The Parent class currently supports only a single child

```kotlin
package org.wiley.kotlin.person

class Parent(_firstName: String, _lastName: String,
             _height: Double = 0.0, _age: Int = 0, _child: Person) :
             Person(_firstName, _lastName, _height, _age) {

    var child: Person = _child
}
```

Person is fine, but Parent currently only supports a single child element. That doesn't model reality, and a collection is a pretty obvious choice. So you could have something like this:

```kotlin
var children: MutableList<Person> = emptyList()
```

Then you could add Person instances to it. But now is where you would stop and think: Is this the right collection to use? Remember, you always have three basic choices: List, Set, Map. List is ordered and can contain duplicates; Set is unordered and cannot; and Map works with key-value pairs.

For now, Map isn't really a candidate. This is just a list of children, via Person objects. So that leaves List and Set. The key differences are order (which *might* matter, if you want to keep your children in order by birthday, or something similar) and duplicates. But that latter property really does carry the day: a set cannot contain duplicates, and a list can. With that in mind, does it make sense to have a list that can have the same child twice? In almost every case, it doesn't.

Keeping that in mind, you can change this to use a set. You also should initialize the set with the child argument that comes into the primary constructor:

```kotlin
var children: MutableSet<Person> = mutableSetOf(_child)
```

Now, in your test class, you can add and remove children in this set quite easily:

```kotlin
val mom = Parent("Cheryl", "Truesby", _child=brian)
mom.children.add(rose)
```

In this case, the mom instance should have two children now—the brian instance and the rose instance:

```
val mom = Parent("Cheryl", "Truesby", _child=brian)
mom.children.add(rose)
println("Cheryl's kids: ${mom.children}")
```

The output is shown here:

```
Cheryl's kids: [Brian Truesby, Rose Bushnell-Truesby]
```

Allow Collections to Be Added to Collection Properties

Now Parent can support multiple children, which is a step forward. But it doesn't support directly adding more than one child at once. That can be handled with a secondary constructor:

```
constructor(_firstName: String, _lastName: String, _children: Set<Person>) :
        this(_firstName, _lastName) {
    children = _children
}
```

Now this code looks OK, but a huge number of problems are buried in just these four lines of code. Compile and you'll get the first line of issues:

```
Error:(10, 13) Kotlin: None of the following functions can be called with the
arguments supplied:
public constructor Parent(_firstName: String, _lastName: String, _height:
Double = ..., _age: Int = ..., _child: Person) defined in
org.wiley.kotlin.person.Parent
public constructor Parent(_firstName: String, _lastName: String, _children:
Set<Person>) defined in org.wiley.kotlin.person.Parent
Error:(11, 20) Kotlin: Type mismatch: inferred type is Set<Person> but
MutableSet<Person> was expected
```

So what's going on here? Well, take it one item at a time. First up, remember that any secondary constructor *must* call the class's primary constructor, which in this case requires a single Person instance for the _child argument. That's immediately a problem, unless you do something rather silly like this:

```
constructor(_firstName: String, _lastName: String, _children: Set<Person>) :
        this(_firstName, _lastName, _child = _children.first()) {
    children = _children
}
```

That technically works, but is not great programming. You're following the rules of Kotlin, but what this really suggests is that your constructors aren't well-planned or executed.

You'll recall that when it comes to inheritance, you want the more general class higher up the inheritance tree, and more specific classes are subclasses. So Person is more general and a superclass, while Parent is more specific and a subclass.

In the same vein, you want your primary constructor to be the broadest, more general constructor, and secondary constructors are more specific versions. That's exactly the *opposite* of what's going on here: the primary constructor takes in a single child (more specific) while the secondary constructor takes in any number of children (including zero, via an empty set).

So switch those up, as shown in Listing 5.9. You'll also find that this makes the calling of the primary constructor from the secondary quite trivial.

LISTING 5.9: Flipping constructors in Parent to function as they should

```
package org.wiley.kotlin.person

class Parent(_firstName: String, _lastName: String,
             _height: Double = 0.0, _age: Int = 0, _children: Set<Person>) :
             Person(_firstName, _lastName, _height, _age) {

    var children: MutableSet<Person> = _children

    constructor(_firstName: String, _lastName: String, _child: Person) :
            this(_firstName, _lastName, _children = setOf(_child))
}
```

In fact, this eliminates the need for any code in the secondary constructor at all. It now just acts as a convenience constructor.

Sets and MutableSets Aren't the Same

But you still have errors:

```
Error:(7, 40) Kotlin: Type mismatch: inferred type is Set<Person> but
MutableSet<Person> was expected
```

What's going on now? The issue here is that you're creating a new `MutableSet<Person>` called `children` and assigning it a `Set<Person>`. But a mutable set and an immutable one are not at all the same, so Kotlin throws an error.

Instead, you'll need to create the new mutable set and then add all the elements from the immutable set to it:

```
var children: MutableSet<Person> = mutableSetOf()

init {
    children.addAll(_children)
}
```

> **NOTE** *It's actually a bit misleading to call the set passed into the primary constructor of* Parent *an immutable set. Rather, it's a set that Kotlin treats as immutable because the type is* Set<Person> *in the constructor's list of arguments. But that's just so that either a mutable or immutable set can be passed in.*
>
> *In reality, it's the need to accept both types of sets at the constructor level that creates this extra code in* init. *That's a tradeoff; rather than only accepting mutable sets in the constructor, all sets are accepted, but then extra code has to be written as a result.*

This all boils down (again) to type safety. Kotlin uses a pretty heavy hand here, by assuming that if you say you're expecting a `Set`, you want a `Set`—and not a `MutableSet`. If you really want to work around that, you've got to do it yourself, which is why you'll need your own `init` block and to call `addAll()`.

There's one additional tweak you can make that cleans this up a bit more. You can actually drop the `init` block and make this assignment when you declare the `children` property:

```
var children: MutableSet<Person> = _children.toMutableSet()
```

Collection Properties Are Just Collections

With this change, you now have the power of collections available through the `children` property of any `Parent` instance. Here's a portion of the ever-growing `PersonTest` using the `filter()` method of children:

```
val mom = Parent("Cheryl", "Truesby", _child=brian)
mom.children.add(rose)
mom.children.add(Person("Barrett", "Truesby"))
println("Cheryl's kids: ${mom.children}")

println(mom.children.filter { it.firstName.startsWith("B") })
```

Obviously, you know that the reason the `children` property has a `filter()` method is because it's a collection, and that's something that collections give you.

For the user, though, it just *feels like* a method that's been defined on `children`. That `children` is a collection is a detail that isn't as important as having a `children` property that is useful. That's the power of collections: lots of methods, and easy to wrap into a custom object.

At this point, you've got a great body of knowledge on classes. In the next chapter, you'll step away from classes (but only briefly!) and learn about generics and how those `Set<Person>` declarations really work.

The Future (in Kotlin) Is Generic

WHAT'S IN THIS CHAPTER?

➤   What is a generic type in Kotlin?

➤   Why are generics needed?

➤   Type projections

➤   Covariance, contravariance, and invariance

GENERICS ALLOW DEFERRING OF A TYPE

Wow. What a mouthful of a chapter heading. It's appropriate, though, as this chapter is going to get into relatively deep waters in terms of Kotlin and the specific topic of generics. This concept is a means of defining a class or a function, and deferring—putting off—the strict definition of a type. As confusing and possibly odd as that may sound, you've already been using generics a *lot*.

Collections Are Generic

As an example, take a look at Listing 6.1. It's the first bits of defining the `ArrayList` class, an implementation of `List` that is the default list returned by `listOf()`. You used this class throughout Chapter 5.

LISTING 6.1: Key bits of ArrayList source code

```
/*
 * Copyright 2010-2018 JetBrains s.r.o. Use of this source code is governed by
the Apache 2.0 license
 * that can be found in the LICENSE file.
 */

package kotlin.collections
```

LISTING 6.1 *(continued)*

```
actual class ArrayList<E> private constructor(
        private var array: Array<E>,
        private var offset: Int,
        private var length: Int,
        private val backing: ArrayList<E>?
) : MutableList<E>, RandomAccess, AbstractMutableCollection<E>() {

    // Lots and lots of code...
}
```

The key line is the declaration of the class:

```
actual class ArrayList<E> private constructor(
```

In particular, look at that odd `ArrayList<E>`. In this instance, `ArrayList` is what is called a *parameterized class*, and `E` is the parameter. That then makes `ArrayList` a generic class. But there's no Kotlin type named `E`, so what gives?

In this case, `ArrayList` is saying to the Kotlin compiler, "When I'm created, a type needs to be specified. That type is the type that I can hold." This is why you might specify that a variable is an `ArrayList` of `String`s (shown first) or of `Band` objects (shown second):

```
var stringList: ArrayList<String>
var bandList: ArrayList<Band>
```

> **WARNING** *If you're creating a program to test this (and you should!), don't forget to import* `org.wiley.kotlin.music.Band`.

In the first case, `E` stands in for `String`; in the second case, it stands in for `Band`.

Parameterized Types Are Available Throughout a Class

Once you've defined a parameter like `E`, it can (and likely should) be used throughout the class. So here's the definition of the `contains()` method on `ArrayList`:

```
override actual fun contains(element: E): Boolean {
```

Note the use of `E` again. This `E` is a parameter that can stand in for a specific type, but it is the *same* `E` as in the class definition. That means that if you create an `ArrayList<Band>` instance, then every time `E` appears in `ArrayList`, it's interpreted as `Band` for that instance.

If you create an instance of `ArrayList<Band>` and try to call `contains()` and supply it a `String`, you're going to get an error. Try this code:

```
var bandList: ArrayList<Band> = ArrayList<Band>()

println(bandList.contains("Not a band"))
```

When you compile it, you'll get this rather oddly worded error:

```
Error:(7, 22) Kotlin: Type inference failed. The value of the type parameter T
should be mentioned in input types (argument types, receiver type or expected
type). Try to specify it explicitly.
```

That's good! That's the static type safety that's so critical to Kotlin. It's also the power of generics.

> **NOTE** *The use of* T *in the error shouldn't be confusing. Different single letters (*E *and* T *are the most common) all are used at various times for parameters. So treat the* T *in the error in the same way that you would if it were* E*.*

Understanding that error is actually a great step toward getting a better understanding of generics. What's going on with that "type inference failed?" It simply means that Kotlin is taking the input argument—in this case, a `String`—and trying to figure out if it can make that object fit the required type. Since E for this `ArrayList` is Band, it rejects that. Kotlin attempts to suggest a resolution: "Try to specify it explicitly." But that doesn't work, because a `String` can't be forced into a `Band`.

This also brings up a good point: sometimes you have to do a little searching online (and re-reading) to figure out a Kotlin error message. That's the case here; it would be nice if it simply said, "Please supply a Band argument." But the more formal message that Kotlin gives is correct: it is trying to infer from the input argument a type that would fit for `contains()`, and couldn't.

Generic: What Exactly Does It Refer To?

It can sometimes be tricky to get programming vocabulary correct. And while the most important thing about being a good programmer or developer is to actually write great code, knowing the terminology is helpful. In the case of generics, the term *generic* refers to a type. Further, the term is usually used to represent the overall type, not a specific instance.

So you might say or hear that in Kotlin, lists are generic. That makes sense; the overall `List` type (as well as other collections) is generic and uses that E (or sometimes T) to keep a parameter open. But, you wouldn't say that you have a generic list and reference a specific variable (like bands). The overall type is generic, but once you instantiate one, you've got something else (even if it's just a list of Any object instances, the broadest possible list instance type).

In addition to generic types, you can have generic functions. These are just what you'd think: functions that are defined in such a way that the types they return are left open to later definition. For instance, suppose you had an `Animal` class that defined a `breed()` method. That method might return a `List<T>` as its return type. Then subclasses would take that generic method and refine the return type (perhaps to `List<Dog>`). You can start to take this pretty far, so keep reading for a lot more on using generics in practical ways.

> **NOTE** *There's actually quite a deep rabbit hole to explore if you want to get into the mechanics of Kotlin and generics, and you'll quickly find yourself learning about Java and its usage of generic types as well. If you are into programming language theory, you might want to check out* `kotlinlang.org/docs/refer-ence/generics.html`, *which takes a much deeper and more theoretical dive into Kotlin generics and their history as it reaches back into Java.*

GENERICS TRY TO INFER A TYPE WHEN POSSIBLE

As with any other language, the more you use Kotlin, the more you'll learn where you can let Kotlin figure things out, and you do less work (which usually means less typing). For example, if you create a new instance of a generic type, Kotlin will do its best to figure out the parameter type that needs to be assigned to that type.

This is called type inference, and it's a very important part of understanding Kotlin, generics, and what can be some very annoying exceptions when trying to compile code.

Kotlin Looks for Matching Types

For example, suppose you have code like this:

```
val brian = Person("Brian", "Truesby", 68.2, 33)
val rose = Person("Rose", "Bushnell", brian)
var personList = mutableListOf(brian, rose)
```

Kotlin assumes that since you've given the `mutableListOf()` method two `Person` instances, you must want a list of `Person`. So it infers that you actually are doing this:

```
var personList = mutableListOf<Person>(brian, rose)
```

Notice that here, the generic `List` is given a parameter type: `Person`. This should make perfect sense. It saves you a step, and is why in earlier chapters you could do things like this:

```
var wordList = mutableListOf("rose", "any", "other", "name", "rose")
var primes = mutableListOf(1, 2, 3, 5, 7, 11)
```

In each case, Kotlin makes type inferences based on the contents of your list. The first is a list of `String` objects and the second is a list of `Int`s.

Kotlin Looks for the Narrowest Type

But Kotlin does more than just look for a matching type. When it's doing type inference, Kotlin looks for the *narrowest* matching type. In other words, it is looking for the most restrictive type that matches all incoming arguments.

Consider the following code:

```
val brian = Person("Brian", "Truesby", 68.2, 33)
val rose = Person("Rose", "Bushnell", brian)
var personList = mutableListOf(brian, rose)
```

Kotlin sees two instances of `Person` and infers that you want a `List<Person>`. That certainly seems reasonable.

Now suppose you have the other extreme:

```
var anyList = mutableListOf("Words", 42, 33.8, true)
```

Here, there really isn't a common type other than the broadest Kotlin type, `Any`. So you're getting a `List<Any>`. While there's nothing wrong with this, it's not going to give you much type safety. You can add literally any type to this list, and you'll have no problems.

Still, in both these cases, the results are somewhat predictable.

Sometimes Type Inference Is Wrong

Consider a case where things might not be so predictable based on type inference:

```
val mom = Parent("Cheryl", "Truesby", _child=brian)
val anotherMom = Parent("Marie", "Bushnell", _child=rose)
val parentList = mutableListOf(mom, anotherMom)
```

Kotlin sees two `Parent` instances and therefore infers that you want a `List<Parent>`. But is that the actual intent? Or do you actually want a `List` of `Person` instances, and it just happens that `mom` and `anotherMom` are `Parent` instances (which are subclasses of `Person`)?

This is where type inference can be problematic. Kotlin doesn't know your intention, which is why you'll get an exception here:

```
val mom = Parent("Cheryl", "Truesby", _child=brian)
val anotherMom = Parent("Marie", "Bushnell", _child=rose)
val parentList = mutableListOf(mom, anotherMom)

parentList.add(brian)
```

The exception is this:

```
Error:(20, 20) Kotlin: Type mismatch: inferred type is Person but Parent
was expected
```

That's correct; you created a `List<Parent>` because you let Kotlin infer your type. If you wanted to allow `Person` instances, you'd need to be explicit:

```
val mom = Parent("Cheryl", "Truesby", _child=brian)
val anotherMom = Parent("Marie", "Bushnell", _child=rose)
val parentList = mutableListOf<Person>(mom, anotherMom)

parentList.add(brian)
```

This code compiles happily, as you've been explicit about your types and now allow `Person` instances in `parentList`.

Don't Assume You Know Object Intent

If this seems a little manufactured, it actually happens more than you might think. Lots of times, you aren't creating an object instance and then creating a collection in the same bit of code, let alone in subsequent lines of code. More often, you are writing a method that takes in an argument—and you don't always know the intent of that object.

You might get a `Parent` that is meant to narrowly be a `Parent`. You might get a `Parent` that is really—at least for the purposes of your usage—just a `Person` and doesn't need to be considered narrowly as a `Parent`. Since you don't know that, you won't always know what type inference will result in.

The better approach is to use types whenever you can and avoid type inference. That ensures that you are in control of types and what Kotlin does to the greatest degree.

Kotlin Doesn't Tell You the Generic Type

One of the most frustrating things about this situation is that you can't print out the type that Kotlin has inferred—or even that you've assigned—to a collection or any other generic type. You can print out the class with `variable::class.simpleName`, but that will only give you `List` or `Band`. It will not give you `List<Person>` or `List<Parent>` or `Band<Rock>`.

This can be frustrating, as the only time you *do* find out the inferred type is when you have an exception:

```
Error:(18, 20) Kotlin: Type mismatch: inferred type is Person but Parent
was expected
```

Here, you can see that the inferred type is `Person`. But there's no other way to get that information easily.

You *can* grab the type from an object in the list, but that's not always accurate. Consider this little bit of code:

```
val brian = Person("Brian", "Truesby", 68.2, 33)
val rose = Person("Rose", "Bushnell", brian)
val mom = Parent("Cheryl", "Truesby", _child=brian)
var inferredList = mutableListOf(mom, brian, rose)
```

If you print out the first item in the list, you're going to get a `Parent`, but items later in the list are `Person`, meaning that the inferred type would be the less specific type, `Person`.

Just Tell Kotlin What You Want!

This is something that's been said before, but if there is anything you take away from this section, it's to simply declare your types explicitly whenever possible. This removes a concern about a narrower type causing Kotlin to type your list or set or collection (or some other generic type) incorrectly. It's also going to improve your code; you're leaving less room for inference (by other people who read your code, not just the Kotlin compiler) and making your intent very clear.

COVARIANCE: A STUDY IN TYPES AND ASSIGNMENT

It's time to get a little further into the weeds with generics and collections. So far, you've basically assumed a method signature for mutable collection methods that looks like this:

```
fun add(element: E): Boolean
```

> **NOTE** *A* method signature *is the more formal term for how a method is defined, including the method's name, return type, any input arguments, and any other modifiers (like* override*). It's typically a term used in more formal programming discussions.*

But that E does more than just accept type E, as you've just seen. It accepts any subtypes of E, as well. So if you have a list of Band objects, making E Band, and RockBand is a subclass of Band, then the add() method for that list takes RockBand instances, as well as Band instances (and any other subclass instance of Band).

This is because Kotlin is happy to accept subclasses in the place of a superclass, as long as the superclass itself is allowed. You've seen this for most of this book by now, so no big surprises.

What about Generic Types?

What you have *not* seen yet is trying to move between various types that are more complex than Band and RockBand or Person and Parent—types that are generic.

Suppose you created an additional subclass of Person called Child. Like Parent, this adds a new constructor that takes in additional information, and a new accessor method. Listing 6.2 shows this new class. You should add it to your own IDE and project in the org.wiley.kotlin.person package.

LISTING 6.2: New Child class for use in type testing and experimentation

```
package org.wiley.kotlin.person

import org.wiley.kotlin.person.Person

class Child(_firstName: String, _lastName: String,
            _height: Double = 0.0, _age: Int = 0, _parents: Set<Person>) :
    Person(_firstName, _lastName, _height, _age) {

    var parents: MutableSet<Person> = mutableSetOf()

    init {
        parents.addAll(_parents)
    }

    constructor(_firstName: String, _lastName: String, _parent: Person) :
        this(_firstName, _lastName, _parents = setOf(_parent))
}
```

This class is really just a counterpart to Parent and doesn't have much new to explore. Now, create a new test class—call it VarianceTest—and create a bunch of new Parent, Child, and Person instances. This class is shown in Listing 6.3.

LISTING 6.3: New test class to look at variance in Kotlin

```
import org.wiley.kotlin.person.Child
import org.wiley.kotlin.person.Parent
import org.wiley.kotlin.person.Person

fun main() {

    // Add some people
    val brian = Person("Brian", "Truesby", 68.2, 33)
    val rose = Person("Rose", "Bushnell", brian)
    val leigh = Person("Leigh", "McLaughlin")

    // Add some parents
    val cheryl = Parent("Cheryl", "Truesby", _child=brian)
    val shirley = Parent("Shirley", "Greathouse", _child=leigh)

    // Add some children
    val quinn = Child("Quinn", "Greathouse", _parent=shirley)
    val laura = Child("Laura", "Jordan", _parent=shirley)

    val momList: List<Parent> = listOf(cheryl, shirley)
    val familyReunion: List<Person> = momList
}
```

So far, you've worked with collections as containers, and not really dealt with them as variables on their own. In other words, you haven't created one variable and assigned a collection to it, and then moved between different collection types. But that's what's going on here.

In a nutshell, there is a list of parents (`momList`, a `List<Parent>`) and a list that would hold lots of people (`familyReunion`, a `List<Person>`). Near the end of the test class, a `momList` of type `Parent` is assigned to a `List` of type `Person`. This is allowed in this case because of two things:

➤ Kotlin collections are *covariant*: a generic class with a subtype of a type `E` can be assigned to a generic class with a type of `E`.

➤ Kotlin has already done the behind-the-scenes work to set this covariance up for you. (More on this shortly.)

> **WARNING** *At this point, your eyes may have just glazed over. That was quite a mouthful and may require you to reread the statement over a few times. That's OK. This next section gets fairly deep into programming theory. That said, take your time, and you'll come out the other side with a better understanding of generics, variance, and programming in general.*

Some Languages Take Extra Work to Be Covariant

In many languages, covariance is not automatic. So in Java, suppose you had this same assignment:

```
List<Parent> momList = new ArrayList(cheryl, shirley);
List<Person> familyReunion = momList;     // ERROR!
```

You'd get an error here. That's because Java's collections are *invariant*. You cannot add a type that has a parameter subtype to a type that has a parameter with that subtype's superclass (or, in fact, anywhere up the inheritance chain).

Kotlin Actually Takes Extra Work to Be Covariant, Too

It's time for a little surprise: Kotlin actually is *not* covariant by default. It's just that Kotlin has already defined its collections as covariant. Here's the actual opening declaration for a `List` in Kotlin:

```
interface List<out E> {
```

Notice the all-important new keyword `out`. That one word is making this collection covariant. It says something like, "methods of this class can only return type E." Kotlin can then take that E and see if it's a subclass of another type (like the type of a list you want to assign these contents to).

So consider this assignment:

```
List<Person> familyReunion = momList;
```

Kotlin knows that `momList` methods will return type `Parent`, because the `out` keyword guarantees that. Kotlin also knows that a `Parent` will work as an input to `familyReunion`, because it's a `List<Person>`, and if the list takes a `Person`, it can take a `Parent`, too. So the covariance here means this code compiles.

Any class, then, that uses `out` is going to be covariant.

Sometimes You Have to Make Explicit What Is Obvious

It's worth taking a quick mental break here to acknowledge something that may have occurred to you: that `out` word is telling Kotlin something that you might expect Kotlin to know on its own. Shouldn't it be obvious that the members of a list of `Parent` objects should be addable to a list of `Person` objects, given that `Parent` inherits from `Person`? It does seem like it.

But this is a case where humans are still required to be smarter—and more deliberate—than the computer. You have to tell Kotlin exactly what you want to be sure it understands that yes, these collections should be covariant. And get ready, because some more unintuitive nuances are about to show up!

Covariant Types Limit the Input Type as Well as the Output Type

You know now that a collection that's covariant—like `List`—is covariant because it tells Kotlin, "I will only return a specific type, E" and for a `List<Parent>`, it's further saying, "That type is Parent." That's great; you can assign a `List<Parent>` to a `List<Person>` because of that declaration.

But covariance has another constraint: covariant types only return the type E, but they also *cannot* accept the type E. That means that you won't be able to add() to a List<Parent> with Parent instances. You can either get Parent instances out, *or* you can put Parent instances in (like a MutableList). But you can't do both.

So, to recap:

➤ *Covariant types*, using the word out, return type E, but cannot take type E as an input parameter to methods. The List<E> type in Kotlin is covariant.

➤ *Invariant types*, which do not use out, can take as input type E, but cannot be cast to types that do not exactly match them. The MutableList<E> types in Kotlin are invariant.

Of course, this makes sense, as the most important thing about a mutable class is its ability to be mutated—which means it needs to take in its type (E). That knocks out covariance for mutable collections.

Covariance Is Really about Making Inheritance Work the Way You Expect

If you think about it, all of this discussion is really making explicit what you likely expect to happen. Covariance allows a class typed as RockBand to work as an input for something that expects a class typed as Band. Similarly, you can accept a List<Parent> as input when you expect a List<Person>. It's what you *expect* to happen based on subclasses and superclasses, and that's important to remember.

Another important way to think about this is that a covariant class is a *producer*. The out keyword tells you that the class will produce the type:

```
interface List<out E> {
```

This list is now a producer of type E. So if you have a List<Person>, the list produces E. Producers are covariant. The cost for being a producer is that the list cannot take as arguments E.

Next up, you're going to see another type of variance: contravariance, or consumers. They are in many ways the mirror image of covariants.

CONTRAVARIANCE: BUILDING CONSUMERS FROM GENERIC TYPES

So you've seen how you can limit the output type of a type, and by doing so, allow for inheritance to work in intuitive ways. But what about mutable types? What about types where you want to focus on what goes into the class, rather than what comes out? That's contravariance.

Contravariance: Limiting What Comes Out Rather Than What Comes In

Covariance limits a generic class's ability to take in certain types, and invariance doesn't limit anything. The net result is that covariance types can be cast to supertypes, and invariant types cannot.

But there's another variance: *contravariance*. Take a look at the `Comparator` interface in Listing 6.4, which will help you see contravariance in action.

LISTING 6.4: The Comparator interface in Kotlin is contravariant

```
package kotlin


public interface Comparator<in T> {
    /**
     * Compares this object with the specified object for order. Returns zero if
this object is equal
     * to the specified [other] object, a negative number if it's less than
[other], or a positive number
     * if it's greater than [other].
     */
    public abstract fun compare(a: T, b: T): Int
}
```

You're going to learn much more about interfaces in the coming chapters, so don't get too hung up on the semantics. Basically, this is a method that classes can implement to compare two objects (of type `T`).

This interface is somewhat unique in that you will often see various implementations using the same type. That's because you often want to compare things differently. Let's say that you wanted to compare different `Person` instances based on last name. You could create a quick `Comparator` implementation like this:

```
val comparePeopleByName = Comparator { first: Person, second: Person ->
    first.lastName.first().toInt() - second.lastName.first().toInt()
}
```

> **NOTE** *Consider this a sneak peek at interfaces and implementing them. It also is a bit of a more advanced way to create an implementation: it's doing that inline, rather than using an entire separate class file.*

Don't get too hung up on the details here. What you *should* see is that this new object implements `Comparator` and overrides `compareTo`. It takes in two instances of `Person`, and compares the lastName properties, returning positive if the first comes before the second, negative if the second comes before the first, and equal if the two start with the same letter.

> **WARNING** *This is a simplistic comparison. Obviously, you can beef this up to check more than the first letter, but it works well for demonstration purposes.*

What's cool is that you can hand a Comparator implementation—like the one you just coded—to a list and sort the list using that Comparator.

Check out Listing 6.5, which adds the implementation of Comparator as well as uses it to sort a list.

LISTING 6.5: The Comparator interface in Kotlin is contravariant

```
import org.wiley.kotlin.person.Child
import org.wiley.kotlin.person.Parent
import org.wiley.kotlin.person.Person

fun main() {

    // Add some people
    val brian = Person("Brian", "Truesby", 68.2, 33)
    val rose = Person("Rose", "Bushnell", brian)
    val leigh = Person("Leigh", "McLaughlin")

    // Add some parents
    val cheryl = Parent("Cheryl", "Truesby", _child=brian)
    val shirley = Parent("Shirley", "Greathouse", _child=leigh)

    // Add some children
    val quinn = Child("Quinn", "Greathouse", _parent=shirley)
    val laura = Child("Laura", "Jordan", _parent=shirley)

    val gary = ParentSubclass("Gary", "Greathouse", _child=leigh)

    val momList: List<Parent> = listOf(cheryl, shirley)
    val familyReunion: List<Person> = momList

    val mutableMomList: MutableList<Parent> = mutableListOf(cheryl, shirley)
    val familyReunion2: List<Person> = momList

    val comparePeopleByName = Comparator { first: Person, second: Person ->
        first.lastName.first().toInt() - second.lastName.first().toInt()
    }

    println(familyReunion.sortedWith(comparePeopleByName))
}
```

The output here will list the instances in `familyReunion` alphabetically:

```
[Shirley Greathouse, Cheryl Truesby]
```

No big deal (other than having just used your first interface!). The names have been sorted using the `Comparator` implementation in `comparePeopleByName`.

Contravariance Works from a Base Class Down to a Subclass

With covariance, you were thinking *up* the inheritance chain:

```
List<Parent> momList = new ArrayList(cheryl, shirley);
for (mom in familyReunion) {
    println(mom.fullName())
}
```

Here, you're passing in a subclass (a `Parent` instance) but you can treat it as the base class. You start specific and become more general. That's covariance. You're producing a certain type, thus the `out` keyword. Everything that comes *out* of `momList` is a `Person`, and so you can operate on those things as if they are `Person` instances.

Contravariance goes the other way. You're thinking about a base class and working down to a subclass, all considering what goes *into* the class:

```
val comparePeopleByName = Comparator { first: Person, second: Person ->
        first.lastName.first().toInt() - second.lastName.first().toInt()
}

println(familyReunion.sortedWith(comparePeopleByName))
```

You are taking in a type specified as `Person`, and so anything that can be considered a `Person` goes *into* `Comparator`. This also means that contravariant classes are *consumers*. They take *in* a certain type. That's why the keyword `in` is used:

```
public interface Comparator<in T> { ... }
```

Contravariant Classes Can't Return a Generic Type

Remember that covariant classes were producers and could only return the generic type specified; they could not take in that type. Contravariant classes are consumers and can take in the specified type; they cannot return that type.

Consider this fictional method signature:

```
public interface Comparator<in T> {
        public abstract fun greatest(a: T, b: T): T
}
```

Would this be legal? No, because `Comparator` is contravariant. It is a consumer, so it can take in type `T`, but it cannot produce type `T`. The `greatest()` method attempts to return type `T`, which is illegal.

This turns out to be an important thing to keep in mind. Classes in Kotlin can be covariant or contravariant. They cannot be both, though. Table 6.1 highlights the differences.

TABLE 6.1: Differences between Covariant and Contravariant

TYPE	KEYWORD	RESPONDS AS	CAN RETURN A GENERIC TYPE?	CAN ACCEPT A GENERIC TYPE?	EXAMPLE
Covariant	out	Producer	Yes	No	List
Contravariant	in	Consumer	No	Yes	Comparator

Does Any of This Really Matter?

If your eyes are rolling back into your head and you're stuck thinking of "contra" as the name of a sidescroller video game, you're in good company. Covariance and contravariance are heavy topics, and you're not going to find a lot of discussion around them at typical programmer gatherings. They might turn up in an advanced session on Kotlin at a conference or in an article on medium.com, but it's nothing that is going to come up in your every-single-day programming life.

Additionally, many a Kotlin programmer (or programmer of Java, or any other inheritance-heavy language) has lived out their days never once worrying about whether to use in or out. You might occasionally run into an error like this:

```
Type parameter T is declared as 'in' but occurs in 'out' position in type T
```

But that's really it. Even then, you could likely fiddle around with your code until it compiles.

So then . . . why all the fuss?

Here's why: if you want to gain a deep understanding of inheritance, you're going to need at least a basic understanding of covariance and contravariance. You should recognize the in and out keywords, because they appear in the Kotlin source code, especially in collections. And yes, sometimes you may want to engage in one of those advanced-level sessions at the dusty end of a conference hall.

UNSAFEVARIANCE: LEARNING THE RULES, THEN BREAKING THEM

As long as you are hanging out in the restricted area of the library, there's one more keyword you should learn, and it's actually what's called an annotation: @UnsafeVariance.

> **NOTE** *An annotation is a form of metadata in Kotlin. It's a way to write something in your code that speaks to the compiler directly, often giving it an instruction.*

The @UnsafeVariance annotation is used like this:

```
override fun lastIndexOf(element: @UnsafeVariance E): Int = indexOfLast { ... }
```

It tells the compiler to suppress warnings related to variance, which is generally a pretty bad idea. It's basically telling the compiler "I've got this" (whether you do or not). It also ends up telling the compiler "Whoever inherits this code later also has this," which is less likely—especially since these sorts of annotations often go unnoticed until four or five hours into a long debugging session.

The good news is that you can probably go most of your Kotlin programming career without resorting to using @UnsafeVariance. The bad news is that even the Kotlin source code occasionally uses this annotation to "make things work."

> **WARNING** *There's not an extended example of using @UnsafeVariance because it's really not a great idea. If you're determined to see it in action, you can Google "Kotlin UnsafeVariance" and you'll find some examples. Just think carefully before you put it into your own code!*

TYPEPROJECTION LETS YOU DEAL WITH BASE CLASSES

So far, you've been looking at what's called *declaration site variance*. That's a fancy term that means you're using the in and out keywords at the class level to make a class either covariant or contravariant.

Variance Can Affect Functions, Not Just Classes

Sometimes you want to operate at a smaller level—such as that of a particular function. For example, suppose you wanted to build a function to copy all the elements from one array into another array. You might code that like this:

```
fun copyArray(sourceArray: Array<Any>, targetArray: Array<Any>) {
    if (sourceArray.size == targetArray.size) {
        for (index in sourceArray.indices)
            targetArray[index] = sourceArray[index]
    } else {
        println("Error! Arrays must be the same size")
    }
}
```

This looks simple enough. As long as the arrays match up in terms of length, the elements are pulled out from one and assigned to the corresponding position in the other. No big deal.

But try running this code with a couple of arrays with actual types:

```
val words: Array<String> = arrayOf("The", "cow", "jumped", "over", "the", "moon")
val numbers: Array<Int> = arrayOf(1, 2, 3, 4, 5, 6)

copyArray(words, numbers)
```

Compile, and you're going to get errors:

```
Error:(41, 15) Kotlin: Type mismatch: inferred type is Array<String> but Array<Any>
was expected
Error:(41, 22) Kotlin: Type mismatch: inferred type is Array<Int> but Array<Any>
was expected
```

So what's happening here? Well, you're about to put your knowledge of covariance and invariance into practice.

First, look at the definition of the `Array` class:

```
class Array<T> { ... }
```

So that's it. It's not covariant, and not contravariant; it's invariant.

> **NOTE** *You'll find it increasingly common to want to see how classes are defined in Kotlin (or any other language). This is easy; just Google "Kotlin source code [class name]" and you'll almost always get a direct link to that class's source code in GitHub.*

That invariance is why you're having a problem with the `copyArray()` function: an array of `String` can't be handed to a function that expects an array of `Any`. For that to work, `Array` would need to be covariant—it would need to take `in` a type and be able to cast it up the inheritance chain.

Thankfully, there's a solution that doesn't involve covariance or contravariance. That's good, as you can't exactly go change the source code for `Array`! It's called *type projection*, and it's a sort of localized version of the declaration site variance you've already seen.

Type Projection Tells Kotlin to Allow Subclasses as Input for a Base Class

Look again at the method signature for `copyArray()` and think about what you need to happen:

```
fun copyArray(sourceArray: Array<Any>, targetArray: Array<Any>) { ... }
```

You can use the `out` keyword here, as you've seen before:

```
fun copyArray(sourceArray: Array<out Any>, targetArray: Array<out Any>) { ... }
```

This is a sort of in-place contravariance. (That's not a technical term as much as a way to think about this paradigm.) It allows the input to both `sourceArray` and `targetArray` to accept not just `Array<Any>` but any subclass of `Any`—so an `Array<String>` or `Array<Int>` would both be acceptable inputs.

If you recompile your code, you'll still get an error—but a different one:

```
Error:(47, 11) Kotlin: Out-projected type 'Array<out Any>' prohibits the use of
'public final operator fun set(index: Int, value: T): Unit defined in kotlin.Array'
```

The first thing to note is that the errors related to passing in an `Array<String>` or `Array<Int>` to `Array<Any>` are gone. That's a step in the right direction.

But there's a new problem. Note that the compiler indicates you now have an "out-projected type." That's exactly right; you've projected a type, allowing for any subclass of the declared type to be accepted. Your function now accepts both arrays you supplied as input.

Producers Can't Consume and Consumers Can't Produce

However, by using type projection, you've created a new problem: you've turned `targetArray` and `sourceArray` into producers. They can now produce `Any` (or subtypes of `Any`). That's not a problem for `sourceArray`, but look at this statement:

```
targetArray[index] = sourceArray[index]
```

This statement is assigning a value from an out `Any` to `targetArray`—which is a producer. As you've already seen in detail, producers cannot consume. So assigning something to `targetArray` is going to generate an error; the `set()` method becomes unusable because you turned it into a consumer via the `out` keyword.

Now, you have a few options here. One thing you *must* do is return `targetArray` to being invariant, so you can assign values to it again:

```
fun copyArray(sourceArray: Array<out Any>, targetArray: Array<Any>) {
```

This is good (even though your code breaks). Now you can assign values from `sourceArray`—which remains contravariant—to `targetArray`. But you're back to an error you've seen before:

```
Error:(41, 22) Kotlin: Type mismatch: inferred type is Array<Int> but Array<Any>
was expected
```

You can't take in for `targetArray` anything now but an `Array<Any>`. So what do you do?

Variance Can't Solve Every Problem

Well, you could change your input array:

```
val words: Array<String> = arrayOf("The", "cow", "jumped", "over", "the", "moon")
val numbers: Array<Any> = arrayOf(1, 2, 3, 4, 5, 6)

copyArray(words, numbers)
```

You're now passing in an `Array<String>` to the `sourceArray` argument, which works because you've functionally made `sourceArray` covariant. You're then passing in exactly what `targetArray` needs—an `Array<Any>`—because `targetArray` is invariant. The code will now compile and work.

The problem, of course, is that this probably isn't really what you actually *want*. The `numbers` variable *isn't* an `Array<Any>` in any true sense; it's an `Array<Int>`. But you're going to have to use it that way if you want to pass it in to `copyArray()`. Ultimately, this is likely a case where you aren't going to get exactly what you want out of a copy operation like this.

If you dig deeper into Kotlin, you'll find that this copying issue shows up a lot with collections. You are going to have a hard time copying one non-typed array into another non-typed array. It's much easier to, for example, take an array that is typed—like `Array<Int>`—and copy additional integers into it. In that case, you don't have two variance issues.

And now you've discovered why variance is more than just about the depths of Kotlin. It has an impact on the code you write in unexpected ways—particularly when dealing with collections.

7

Flying through Control Structures

➤ Using `if` and `else` for conditional code execution

➤ Building expressions with `when`

➤ Selecting from multiple options for execution

➤ Iterating over collections with `for`

➤ Looping until a condition is true with `while`

➤ Adding `do` to `while` for initial execution

➤ Using `break` and `continue` to control flow

CONTROL STRUCTURES ARE THE BREAD AND BUTTER OF PROGRAMMING

Whether you are relatively new to programming or have been writing Java for 20 years, and Kotlin is just an exercise in expanding your programming breadth, you've almost certainly worked with a lot of control structures. In fact, you've already seen these a number of times in this book: every time you've used an if statement or a for, you've been using a control structure.

In this chapter, you'll go beyond passing familiarity with these in Kotlin and dig more deeply into control structures and loops. The obvious place to start is with the most common: if, when, for, and while.

> **NOTE** *Unfortunately, control structures are not particularly exciting. They're important, though, so this chapter strikes a balance: it moves through control structures as fast as possible, but is going to try to unearth some detail you might normally miss or skip.*

IF AND ELSE: THE GREAT DECISION POINT

There's nothing particularly mystical or magical about `if`. It works just like you'd think. Listing 7.1 shows a very simple (*very* simple) program that picks a random number and lets you guess that number.

LISTING 7.1: A very basic usage of if and else

```
import kotlin.random.Random

fun main() {
    var theNumber = Random.nextInt(0, 1000)

    print("Guess my number: ")
    val guess = readLine()!!

    if (guess.toInt() == theNumber) {
        println("You got it!")
    } else {
        println("Sorry, you missed it.")
    }
}
```

A number is taken in through `readLine()`, which actually will return a `String`. That `String` is converted to an `Int`, compared with the randomly generated number in `theNumber`, and then `if` is used to determine if there's a match. If there is a match, a success method is printed. If there's not, the `else` control structure picks up the flow and prints out a consolation method.

!! Ensures Non-Nullable Values

A slight diversion is in order here to treat the odd `!!` construct you may have noticed. If you looked up the definition of `readLine()`, you'd see that it's defined to return `T?`. It returns a parameterized type (where in typical usage, `T` will be `String`), and the `?` indicates that null can also be returned.

If you remove the `!!` so your code looks like this, and compile, you'll get an error:

```
print("Guess my number: ")
val guess = readLine()

if (guess.toInt() == theNumber) {...}
```

The error will indicate a problem with your `toInt()` call:

```
Error:(15, 14) Kotlin: Only safe (?.) or non-null asserted (!!.) calls are allowed
on a nullable receiver of type String?
```

The issue is that `toInt()` will error out if it gets a null value, as will any method called on `guess`. Kotlin recognizes that `guess` could be null—because `readLine()` returns `T?`—and so it throws an error.

To avoid that, you need to ensure that `guess` is not null. You have a number of ways to do that, but one of the easiest is to use `!!`. That operator converts a nullable value into a non-nullable one, and if the input to it (in this case, `readLine()`) is null, it throws a `NullPointerException`.

With that addition, as originally shown in Listing 7.1, Kotlin sees that there's no way for `guess` to have a null value by the time `toInt()` is called. The compiler is happy, and the error goes away.

> **WARNING** *While adding* `!!` *was an easy solution here, it's also pretty brute force. Your program is going to exit abruptly with a stack trace in the event of a null entry. This works here because it's a trivial test program that's already pretty basic, but would almost certainly be a poor idea in a production system or application. In those cases, you'd want to handle nulls gracefully and likely provide users a chance to re-enter their value. The easiest solution would be to just* `return` *and end the program.*

Control Structures Affect the Flow of Your Code

`if` and its companion `else` are simple to use, but critical because they do something fundamental to your program: they affect the flow of execution. That's why these constructs are called *control structures*: they allow you to control that flow.

Figure 7.1 shows the sort of thinking you'll need to adopt as you begin to work more with control structures in this chapter and in your own code.

This initial example is pretty simple. Code begins to execute and then when it hits the `if`, flow splits. Execution goes to either the `if` path or the `else` path, as shown in Figure 7.2. Finally, both blocks end in the same place, and flow merges (just long enough to exit the program, in this case).

Now this is pretty basic here, and you probably didn't need to get out a pencil or marker and diagram the flow. You can see it pretty instantly.

However, as your control structures get more complex, as you use many of them, and you start to nest them, it will become trickier to map control flow. Add to this that many times you will then call methods within your `if`, `when`, `else`, `for`, and `while` blocks, and it can be quite difficult to follow programming logic unless you take a little extra time.

This is very important, though. As you'll see throughout this chapter, knowing exactly what's going on in your code is the key to writing good code, and ensuring that flow goes *exactly* the way you want it to is key to writing great code.

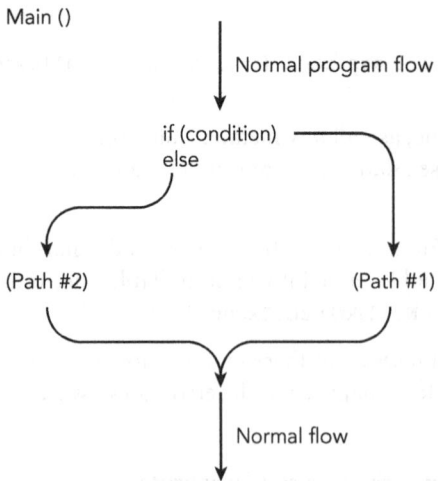

FIGURE 7.1: An if statement adjusts how your code is executed and the flow of code through that execution.

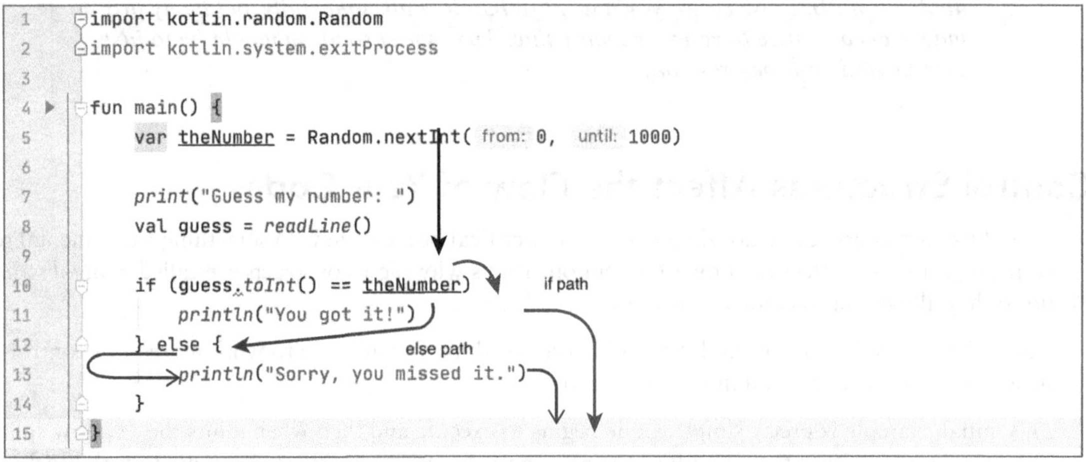

```kotlin
1   import kotlin.random.Random
2   import kotlin.system.exitProcess
3
4   fun main() {
5       var theNumber = Random.nextInt( from: 0,   until: 1000)
6
7       print("Guess my number: ")
8       val guess = readLine()
9
10      if (guess.toInt() == theNumber)          if path
11          println("You got it!")
12      } else {              else path
13          println("Sorry, you missed it.")
14      }
15  }
```

FIGURE 7.2: Program execution has two different flows in this example.

if and else Follow a Basic Structure

Every if/else block follows the same basic structure:

```kotlin
if (EXPRESSION) {
    // code to run if EXPRESSION is true
} else {
    // code to run if EXPRESSION is not true
}
```

You've seen this in action already:

```
if (guess.toInt() == theNumber) {
    println("You got it!")
} else {
    println("Sorry, you missed it.")
}
```

Of course, you don't always have to have an `else` component:

```
if (guess.toInt() == theNumber) {
    println("You got it!")
}
```

Generally, you'll use `if`/`else` if you want to take a different step based on the EXPRESSION. You'll only use an `if` in situations where you want to take an *additional* step based on the EXPRESSION.

The curly braces are also optional, provided that you put your entire statement on a single line. So this is perfectly legal:

```
if (guess.toInt() == theNumber) println("You got it!") else println("Sorry.")
```

It's not nearly as readable, though, so in most cases, it's really preferred to break up your single line into parts and reintroduce those curly braces. Your code will be cleaner and easier to read.

Expressions and if Statements

The most important thing about your `if` statement is making sure you write a clear expression (EXPRESSION in the structure shown previously). That expression must evaluate to either true or false.

> **NOTE** EXPRESSION *here is sometimes referred to as a* CONDITION, *or a conditional statement (or even a conditional expression).*

But that's actually not the only expression involved here. The `if` statement itself is an expression that evaluates to a code block—the things within your curly braces. So take this `if` you've already seen:

```
if (guess.toInt() == theNumber) {
    println("You got it!")
} else {
    println("Sorry, you missed it.")
}
```

If the expression is true—if guess.toInt() == theNumber—then the entire `if` statement here evaluates to:

```
println("You got it!")
```

If the conditional expression is *not* true, then the statement evaluates to:

```
println("Sorry, you missed it.")
```

Use the Results of an if Statement Directly

You can use this to your advantage by actually assigning the result of a statement to a variable, or otherwise using it directly. So here's a sample `if` statement that assigns a value based on an EXPRESSION:

```
var response = StringBuilder("The number you guessed is ")
if (guess.toInt() == theNumber) {
    response.append("right!")
} else {
    response.append("wrong!")
}

println(response.toString())
```

> **NOTE** `StringBuilder` *is just what it sounds like—a way to build up strings easily.*

This is just a variation on what you already have done, so nothing very exciting here. But consider that you can put whatever you want within the `if` and `else` block, and that something is returned from the `if` statement's evaluation. So suppose you had an `if` statement like this:

```
val evenOrOdd = "even"
if (evenOrOdd == "even") {
    6
} else {
    9
}
```

In this case, the `if` statement would evaluate to 6. If `evenOrOdd` were set to anything other than "even," the `if` statement would evaluate to 9. So with this in hand, you can actually rewrite the earlier statement that used `StringBuilder`:

```
var response = StringBuilder("The number you guessed is ")

response.append(if (guess.toInt() == theNumber) "right!" else "wrong!")

println(response.toString())
```

What exactly is going on here? It looks a bit odd, but break this down (in particular, the second line). First, take the `if` statement:

```
if (guess.toInt() == theNumber) "right!" else "wrong!"
```

Now, just for clarity, break this further into blocks:

```
if (guess.toInt() == theNumber) {
  "right!"
} else {
 "wrong!"
}
```

This is something that looks quite familiar. There's a conditional expression that evaluates to either true or false. No big deal. And then there is a bit of code for if that expression is true and another bit for if it's false. Also no big deal.

The oddity, though, is that the code in each block isn't a statement that runs, but a simple value: either "right!" or "wrong!" But remember, the entire `if` statement is an expression that is evaluated; the whole thing will result in *either* the string "right!" or the string "wrong!" Then, that string will be handed off to the `append()` method:

```
response.append(if (guess.toInt() == theNumber) "right!" else "wrong!")
```

That makes perfect sense, because the `append()` method of `StringBuilder` takes a `String` as input. That value is appended to `response`'s existing content, and then output.

Kotlin Has No Ternary Operator

If you've used other programming languages like Java, you've seen the equivalent of this sort of behavior and control flow via a ternary operator. This is an operator that does just what the `if/else` statement you've just seen does, but has slightly different syntax.

For example, here's that operator in use in Java:

```
response.append((guess.toInt() == theNumber) ? "right!" : "wrong!")
```

The form actually looks similar to what you've already seen:

```
[EXPRESSION] ? [CODE TO EVALUATE IF TRUE] : [CODE TO EVALUATE IF FALSE]
```

Of course, you could pretty easily map this to an `if` statement:

```
If [EXPRESSION] then [CODE TO EVALUTE IF TRUE] else [CODE TO EVALUATE IF FALSE]
```

That mapping—and how clear it is—led to Kotlin *not* implementing a specific ternary operator.

A Block Evaluates to the Last Statement in That Block

Consider these slightly longer bits of code in both portions of an if statement:

```
if (guess.toInt() == theNumber) {
    response.append("right!")
    true
} else {
    response.append("false!")
    false
}
```

In each block, the expression evaluates to the *last* statement. So the top block would evaluate to true and the bottom block to false.

That means you can start to get a little cleverer about using the result of an `if` statement. Take a look at this code:

```
print("The number you guessed is ")
var hasRightNumber = if (guess.toInt() == theNumber) {
        println("right!")
```

```
            true
        } else {
            println("wrong.")
        false
        }
```

```
    println(hasRightNumber)
```

In this case, the `if` statement is still being used to calculate a return value: either true or false. But it's also doing some printing. However, you can't change this order. So this code would not work:

```
    print("The number you guessed is ")
    var hasRightNumber = if (guess.toInt() == theNumber) {
            true
            println("right!")
        } else {
            false
            println("wrong.")
        }
```

```
    println(hasRightNumber)
```

You'd actually return the evaluation of the `println()` in either case, because that's the last line of the block being evaluated.

if Statements That Are Assigned Must Have else Blocks

One last detail regarding using an `if` to assign a value: when used that way, `if` statements *must* have `else` blocks. That's because an assignment must have something that's being assigned, and the result of an `if` statement with a false conditional is the `else` block . . . which must exist. If it doesn't, there's nothing to assign, and the whole statement is invalid.

WHEN IS KOTLIN'S VERSION OF SWITCH

Most languages have the concept of a switch. A switch is a statement that also manages control flow but provides for multiple options beyond the simple true/false of an `if` statement. In Kotlin, this is handled through the `when` control structure:

```
    var anotherNumber = Random.nextInt(0, 5)

    when (anotherNumber) {
        0 -> println("the number is 0")
        1 -> println("the number is 1")
        2 -> println("the number is 2")
        3 -> println("the number is 3")
        4 -> println("the number is 4")
    }
```

> **NOTE** *It's a little odd that Kotlin didn't use* switch, *as that's a well-understood keyword. It feels a bit like this was a change just to be different. Regardless, it's an easy one to make.*

When you see a when statement, you should read it as "When anotherName is . . ." for the first line:

```
when (anotherNumber) {
```

Then, for each successive line, read it as "0, then do . . ." or "1, then do . . .":

```
0 -> println("the number is 0")
1 -> println("the number is 1")
// etc
```

In other words, each line is a specific criterion, that if met, the code after -> runs.

Each Comparison or Condition Is a Code Block

You can also use curly braces to create blocks for each case:

```
when (anotherNumber) {
    0 -> {
        println("the number is 0")
    }
    1 -> println("the number is 1")
    2 -> println("the number is 2")
    3 -> println("the number is 3")
    4 -> println("the number is 4")
}
```

> **NOTE** *Each case in Kotlin is also sometimes called a* branch. *The terms are interchangeable.*

And of course, you can have multiple lines within those blocks:

```
var even = false
when (anotherNumber) {
    0 -> {
        println("the number is 0")
        even = true
    }
    1 -> println("the number is 1")
    2 -> {
        println("the number is 2")
        even = true
    }
    3 -> println("the number is 3")
    4 -> {
        println("the number is 4")
        even = true
    }
}
```

Each block can do multiple things, and just as with if, the block will evaluate to the last statement in the block.

Handle Everything Else with an else Block

You can use an else with when as well. In this case, else catches anything that wasn't explicitly matched prior to the else:

```
var anotherNumber = Random.nextInt(0, 1000)

var even = false
when (anotherNumber) {
    0 -> {
        println("the number is 0")
        even = true
    }
    1 -> println("the number is 1")
    2 -> {
        println("the number is 2")
        even = true
    }
    3 -> println("the number is 3")
    4 -> {
        println("the number is 4")
        even = true
    }
    else -> println("The number is 5 or greater")
}
```

In this example, anotherNumber can now run up to 1000, and the else handles anything greater than 4. And else can also have its own code block:

```
var anotherNumber = Random.nextInt(0, 1000)

var even = false
when (anotherNumber) {
    0 -> {
        println("the number is 0")
        even = true
    }
    1 -> println("the number is 1")
    2 -> {
        println("the number is 2")
        even = true
    }
    3 -> println("the number is 3")
    4 -> {
        println("the number is 4")
        even = true
    }
    else -> {
        println("The number is 5 or greater")
        even = (anotherNumber % 2 == 0)
    }
}

println("The actual number was $anotherNumber")
```

> **NOTE** *The* % *operator means modulo. It returns the remainder from a division operation. In this case,* anotherNumber % 2 *means that* anotherNumber *is divided by two, and the remainder is returned. If it's 0, the number was evenly divisible by two, and therefore even. If there is a remainder, the number is odd.*

Each Branch Can Support a Range

when is often used on numbers, and branches can handle more than a single value. Here's an example that handles multiple numbers in each branch:

```
when (anotherNumber) {
    0, 1, 2, 3, 4 -> {
        println("the number is less than 5")
    }
    else -> {
        println("The number is 5 or greater")
        even = (anotherNumber % 2 == 0)
    }
}
```

You can also use the in keyword to condense this syntax:

```
when (anotherNumber) {
    in 0..4 -> {
        println("the number is less than 5")
    }
    else -> {
        println("The number is 5 or greater")
        even = (anotherNumber % 2 == 0)
    }
}
```

You just provide in with a lower and upper bound with two dots (..) between them. Also note that in is inclusive of the lower and upper bounds, so in 0..4 will be true for 0, 1, 2, 3, and 4.

You can also negate a range with !, resulting in !in:

```
when (anotherNumber) {
    !in 0..4 -> {
        println("the number is greater than 5")
    }
    else -> {
        println("The number is 4 or less")
        even = (anotherNumber % 2 == 0)
    }
}
```

Each Branch Usually Has a Partial Expression

It's possible to use a when statement as a sort of super-if statement. Each branch condition can be a statement that evaluates to a Boolean value, which is what you've already seen. However, the rules are a bit different than how you've written if statements.

Here's an example opening statement for a when that includes a value on the first line:

```
when (anotherNumber) {
```

The value of anotherNumber is the value being compared. But the statement at the beginning of each branch is actually not a *complete* expression. For instance:

```
0, 1, 2, 3, 4 -> {
        println("the number is less than 5")
    }
in 5..100 {
        println("the number is between 5 and 100")
    }
```

Here, the bits 0, 1, 2, 3, 4 and in 5..100 do not stand on their own. Instead, they combine with the opening value to create a complete statement. So you take:

```
when (anotherNumber) {
```

and that anotherNumber becomes a prefix to each partial expression for each branch. If there is no operator supplied, a comparison is made:

```
anotherNumber == 0
```

If multiple numbers and no operator are supplied, multiple equalities are checked for with an "or" combining the statements:

```
(anotherNumber == 0) OR (anotherNumber == 1) OR (anotherNumber == 2) ...
```

And finally, if an operator is applied, you get something like this:

```
anotherNumber in 5..100
```

This combination expression is evaluated as a Boolean. If the result is true, then the branch is executed. If it's false, the branch is skipped, and the next branch is checked.

If no branches match, and there is an else, then the else branch is executed. If no branches match and there is *not* an else, then no branches are executed, and the flow returns to the code directly after the when.

You have another option, though. You can leave off the variable or expression in the first line of your when, and then include complete Boolean statements for each branch. Here's a case where a when is turned into an if:

```
when {
    (anotherNumber % 2 == 0) -> println("The number is even!")
    else -> println("The number is odd!")
}
```

This isn't particularly helpful, as you could do this with an `if`. It does become interesting when you're dealing with multiple branches, though:

```
when {
    (anotherNumber % 2 == 0) -> println("The number is divisible by 2!")
    (anotherNumber % 3 == 0) -> println("The number is divisible by 3!")
    (anotherNumber % 5 == 0) -> println("The number is divisible by 5!")
    (anotherNumber % 7 == 0) -> println("The number is divisible by 7!")
    (anotherNumber % 11 == 0) -> println("The number is divisible by 11!")
    else -> println("This thing is hardly divisible by anything!")
}
```

This particular example isn't that exciting, but it does show that you can handle a multiple-branch situation easily. To do this with `if`, you'd need multiple `if/else` statements chained together:

```
if (anotherNumber % 2 == 0) {
    println("The number is divisible by 2!")
} else if (anotherNumber % 3 == 0) {
    println("The number is divisible by 3!")
} else if (anotherNumber % 5 == 0) {
    println("The number is divisible by 5!")
} else if (anotherNumber % 7 == 0) {
    println("The number is divisible by 7!")
} else if (anotherNumber % 11 == 0) {
    println("The number is divisible by 11!")
} else {
    println("This thing is hardly divisible by anything!")
}
```

> **NOTE** *Kotlin doesn't have any idea of an* `elsif` *or* `elseif`, *as some languages do. You can simply use* `else` *and then* `if`, *successively.*

Branch Conditions Are Checked Sequentially

The Kotlin runtime will check each branch one at a time, in the order that they appear in your code. Additionally, *only one branch* will execute. So in the previous code, if anotherNumber is 6, it would actually be divisible evenly by 2 and 3, but only the branch with the condition (anotherNumber % 2 == 0) will execute. That's the first branch that matches, and therefore the *only* branch that executes.

Most of the time, you'll naturally write conditions that are mutually exclusive. However, you may want to be intentional about that. You should also note that once a condition completes, the when block is exited. This is different from some languages that require an explicit call to exit a switch or case statement, like the break keyword in Java.

Branch Conditions Are Just Expressions

You've already seen a few variations of different expressions for branches. With a variable in the initial line of when, you can have partial expressions for each branch. If you drop the variable on the initial line, you can have complete expressions for each branch.

Some other variations exist, too. First, on the initial line of your when, you can actually perform computation:

```
when (anotherNumber % 2) {
    0 -> println("The number is even!")
    else -> println("The number is odd!")
}
```

There's not that much surprising here. Kotlin evaluates (anotherNumber % 2) and the result is combined with the partial expressions in each branch to determine which to follow. And you can use functions in that first line as well:

```
when (mom.children.size % 2) {
    0 -> println("The kids can pair up!")
    else -> println("Uh oh, someone will be left out!")
}
```

You can have function calls in branch conditions, too:

```
when {
    canPlayBasketball(mom.children.size) -> println("You can play a good game
of hoop!")
    canPlayTag(mom.children.size) -> "Plenty for tag!"
    else -> println("Uh oh, no good game ideas!")
}
```

> **WARNING** *If you're following along, this code isn't actually working code, as there are no* canPlayBasketball() *or* canPlayTag() *methods defined anywhere. It's purely for demonstration purposes.*

All of these are just expressions, whether in the initial line of a when or in a branch; whether partial or complete.

When Can Be Evaluated as a Statement, Too

Remember that an if statement can actually be evaluated as a whole:

```
print("The number you guessed is ")
var hasRightNumber = if (guess.toInt() == theNumber) {
    println("right!")
    true
} else {
    println("wrong.")
    false
}

println(hasRightNumber)
```

when has the same capabilities:

```
fun isPositiveInteger(theValue: Any): Boolean =
    when {
        (theValue !is Int) && (theValue !is Byte) &&
                (theValue !is Short) && (theValue !is Long) -> false
        (theValue is Int) -> (theValue as Int > 0)
        (theValue is Short) -> (theValue as Short > 0)
        (theValue is Byte) -> (theValue as Byte > 0)
        (theValue is Long) -> (theValue as Long > 0)
        else -> false
    }
```

Here, a function is defined as a when statement, and like if, each branch returns the last statement.

The first condition is a multi-condition: it checks to see if the value (theValue) is not one of a set of types. If it *is* one of those types, it is converted in the appropriate expression (as an Int, or Short, or Byte, or Long).

But remember, if no match is found, the else is run. That means you could remove that first condition, leaving just this:

```
when {
        (theValue is Int) -> (theValue as Int > 0)
    (theValue is Short) -> (theValue as Short > 0)
    (theValue is Byte) -> (theValue as Byte > 0)
    (theValue is Long) -> (theValue as Long > 0)
    else -> false
```

Now, if no type match occurs, you get the same false returned value, without the clutter of that first conditional.

> **NOTE** *This is a slightly new syntax for functions. Rather than using an open and closing set of curly braces for the function definition, it's entirely represented by an = and then a single statement for the function.*

FOR IS FOR LOOPING

The for control structure is likely the second most common structure—with if being the most common—and almost certainly familiar to you if you've done any programming at all. You've already seen it in action with collections, but of course that's not its only usage.

for, at its simplest, is a loop that goes through a set number of things. "Thing" here is meant to be vague, because a for loop can traverse almost anything—as long as there is a starting point and an ending point.

For in Kotlin Requires an Iterator

If you've been programming for long, you've likely seen simple `for` loops that provide some sort of index or counter. Here's an example in Java:

```java
for (int i = 0; i < 5; i++) {
    System.out.println(i);
}
```

And here's what that would look like in JavaScript:

```javascript
for (i = 0; i < 5; i++) {
    text += i + "<br>";
}
```

Kotlin is a bit different. Here's that same loop in Kotlin:

```kotlin
for (i in 0..4) {
    println(i)
}
```

This may not look like a big change, but some important differences are lurking here.

First, Kotlin doesn't support the somewhat typical `for` format of:

```
for (statement 1; statement 2; statement 3) {
    // code block to be executed
}
```

In this format (common to Java, JavaScript, and many other languages), `statement 1` is typically an assignment, `statement 2` is typically a condition that must be true to execute the code block, and `statement 3` is executed after the loop runs each time.

Kotlin, though, requires an iterator, which really does all this to your code internally. Here's the Kotlin code again:

```kotlin
for (i in 0..4) {
    println(i)
}
```

In this code, `0..4` actually creates an iterator on the fly. In other words, this is somewhat equivalent to the following code:

```kotlin
val numbers = listOf(0, 1, 2, 3, 4)
val numbersIterator = numbers.iterator()
```

You've already seen how to use this code in a collection with `while`; it looks like this:

```kotlin
val numbers = listOf(0, 1, 2, 3, 4)
val numbersIterator = numbers.iterator()
while (numbersIterator.hasNext()) {
    println(numbersIterator.next())
}
```

> **NOTE** *We'll come back to* while *in a bit. For now, we're just passing through.*

You can then condense this a bit using `for`:

```
val numbers = listOf(0, 1, 2, 3, 4)
for (item in numbers) {
    println(item)
}
```

This is old hat by now. That's actually great; once you realize that Kotlin is really just doing the same thing internally, `for` loops become something you've already got down.

You Do Less, Kotlin Does More

So when you write this code:

```
for (i in 0..4) {
    println(i)
}
```

Kotlin actually turns `0..4` into a collection, and `i` into an iterator over that collection. You could almost see Kotlin converting this code into something like this:

```
for (i in listOf(0, 1, 2, 3, 4)) {
    println(i)
}
```

Easy enough, right? Then, each time through the loop, `i` is equal to the next item in the collection.

> **WARNING** *Kotlin doesn't quite do this, although it's close. Honestly, for now, getting into the bytecode generated isn't that useful. Just focus on the key idea that Kotlin turns your range into a collection and then creates an iterator on that collection, and you have what you need.*

For Has Requirements for Iteration

Now that you've seen `for` in action, it's worth being a little more specific about the requirements for what can be iterated over. `for` can iterate over anything that provides an iterator, which means it requires:

➤ Something (like a range) that has as a member or function `iterator()`, and:

  ➤ That iterator has a `next()` function, and

  ➤ That iterator has a `hasNext()` function that returns a Boolean value.

➤ That something doesn't have to be a collection, but usually is or can be turned into one (like a range, for instance, `0..4`) by Kotlin.

As long as all of these requirements are met, you're able to use a `for` loop.

You Can Grab Indices Instead of Objects with for

One apparent limitation of using `for` via an iterator is that you're generally going to be going over each item in the collection created. On top of that, you're just getting the item itself, rather than an index.

So with Java, you'd have code like this:

```
int ar[] = { 1, 2, 3, 4, 5, 6, 7, 8 };
int i, x;

// iterating over an array
for (i = 0; i < ar.length; i++) {

    // accessing each element of array
    x = ar[i];
    System.out.print(x + " ");
}
```

Each time through the loop, `i` would be an index, and that index also serves as a loop counter. But you can do the equivalent in Kotlin by grabbing the indices from a collection:

```
int ar[] = { 1, 2, 3, 4, 5, 6, 7, 8 };
int i, x;

for (i in ar.indices) {
    x = ar[i]
    println("${x} ")
}
```

Instead of directly iterating over each item, the `indices` property is used—which itself returns an iterator—and each time through the loop, `i` is an index rather than the item at that index.

In the preceding example, it's a bit useless, as the code would be better to just grab each object in the array. But you might want to print out the index of each item:

```
for (i in ar.indices) {
    x = ar[i]
    println("${x} is at index ${i}")
}
```

One word of warning, though: it is *not* safe to assume that the collection of values returned by `indices` is identical to a counter starting at 0 or 1. If you want a strict counter, you'll likely need to separate that from the indices or objects over which you're iterating:

```
var counter: Int = 1
for (i in ar.indices) {
    x = ar[i]
    println("Item ${counter}: ${x} is at index ${i}")
    counter++
}
```

> **NOTE** *This is a simple example of why using indices as a proxy for a counter isn't always safe.* indices *in this example will run from 0 to 7, while your counter is running from 1 to 8. Even if you think to add 1 to each value for* i, *there are still times where this is unreliable.*

This is a bit clunky, admittedly. However, it turns out that the vast majority of the time, you actually don't want a counter; you just want those actual indices, or perhaps an overall count. You can get both of those already, so the preceding code is rarely necessary.

Listing 7.2 combines all these examples into a single code listing you can use to verify your own work.

LISTING 7.2: All the code snippets so far, in one sample program

```kotlin
import kotlin.random.Random

fun main() {
    var theNumber = Random.nextInt(0, 1000)

    print("Guess my number: ")
    val guess = readLine()!!

    /*
    if (guess.toInt() == theNumber) {
        println("You got it!")
    } else {
        println("Sorry, you missed it.")
    }
     */

    print("The number you guessed is ")
    var hasRightNumber = if (guess.toInt() == theNumber) {
            println("right!")
            true
        } else {
            println("wrong.")
        false
        }

    println(hasRightNumber)

    var anotherNumber = Random.nextInt(0, 1000)

    var even = false
    when {
```

LISTING 7.2 *(continued)*

```
        (anotherNumber % 2 == 0) -> println("The number is even!")
        else -> println("The number is odd!")
}

when {
        (anotherNumber % 2 == 0) -> println("The number is divisible by 2!")
        (anotherNumber % 3 == 0) -> println("The number is divisible by 3!")
        (anotherNumber % 5 == 0) -> println("The number is divisible by 5!")
        (anotherNumber % 7 == 0) -> println("The number is divisible by 7!")
        (anotherNumber % 11 == 0) -> println("The number is divisible by 11!")
        else -> println("This thing is hardly divisible by anything!")
}

if (anotherNumber % 2 == 0) {
    println("The number is divisible by 2!")
} else if (anotherNumber % 3 == 0) {
    println("The number is divisible by 3!")
} else if (anotherNumber % 5 == 0) {
    println("The number is divisible by 5!")
} else if (anotherNumber % 7 == 0) {
    println("The number is divisible by 7!")
} else if (anotherNumber % 11 == 0) {
    println("The number is divisible by 11!")
} else {
    println("This thing is hardly divisible by anything!")
}

println("The actual number was $anotherNumber")

var foo: Long = 23

println(isPositiveInteger(foo))

val numbers = listOf(0, 1, 2, 3, 4)
for (item in numbers) {
    println(item)
}

val ar = intArrayOf(1, 2, 3, 4, 5, 6, 7, 8)
var i: Int
var x: Int

var counter: Int = 1
for (i in ar.indices) {
    x = ar[i]
    println("Item ${counter}: ${x} is at index ${i}")
    counter++
}
}
```

USE WHILE TO EXECUTE UNTIL A CONDITION IS FALSE

So far, you've seen that you can loop over a collection easily, and you can also loop over anything that can be turned into an iterator. But sometimes you want to loop as long as a particular condition is true. Put another way, you want to loop until that condition is false.

This is where `while` comes into play. Like `if` and `else`, `while` operates very similarly to other languages, so you may find this to be routine and particularly easy to pick up.

While Is All about a Boolean Condition

If you have some time on your hands, write and then run this code:

```
while (true) {
    // this will loop endlessly!
}
```

Although this is useless, and an infinite loop, it also illustrates the simplest form of a `while` loop. You have a condition, and as long as that condition is true, the body of the loop will execute:

```
while (CONDITION) {
    // this will loop endlessly!
}
```

`CONDITION` must evaluate to a Boolean here. Additionally, that `CONDITION` is the way that the loop is exited.

> **NOTE** *You have a few other ways to get out of a loop using* `break`, *which are covered a bit later in this chapter.*

What this effectively means is that you should be potentially changing the components of CONDI-TION within the loop body. Otherwise, `CONDITION` will continue to evaluate to true, and you'll get an infinite loop.

At its simplest, you can create a counter out of a `while`:

```
var x: Int = 1
while (x < 6) {
    println(x)
    x++
}
```

> **WARNING** *If you've been following along, you may need to make a few tweaks to your code to get this to run. First, if you've already declared x earlier, remove the declaration line and replace it with* x = 1. *Second, if you compiled in the infinite loop from the previous code, make sure you remove that, or you'll never get your code to stop, let alone get to anything you've added after the infinite loop.*

This is a little silly because you can do the same thing with `for`, and do it in a way that is both cleaner and easier to read:

```
for (x in 1..5) {
    println(x)
}
```

In fact, it turns out that it's pretty hard to do much of interest with `while` except to copy a `for` loop—unless you use some other control structures as well.

One of the more realistic examples for `while` that stays pretty simple is to pull a value from some other piece of code, and then use that value as part of your condition:

```
var condition: Boolean = Random.nextBoolean()
while (condition) {
    println("Still true!")
    condition = Random.nextBoolean()
}
println("Not true any more!")
```

`Random.nextBoolean()` is a simple helper function that returns a random true or false value. This `while` loop grabs that value, and then loops as long as `condition` is true. Note that within the `while` loop, the next Boolean is pulled each time. Don't forget this, or `condition` will never change, and you'll have another infinite loop!

> **NOTE** *There is one way to avoid this infinite loop: set* condition *to false before it begins. The loop will never execute.*

This code *still* isn't particularly exciting, but it does give you a look at how most `while` loops look: they pull a value from another piece of code, or calculate a value, and then plug that into the condition.

A Wrinkle in while: Multiple Operators, One Variable

One note that bears a bit of explanation: the code so far—and statements about how `while` works— is making a pretty safe assumption that nothing outside your code is changing the values used in the `while` loop's condition. However, it's possible that this isn't the case.

Consider a variable called `condition` that is either true or false, and that value is used as the condition in a `while` loop. If that variable is changed by any other code, it would potentially affect your `while` loop. But how could that happen if you don't change `condition` in your own code?

The simplest case would be to pass `condition` to a function that then changes `condition` within the function. That would look like this:

```
var condition: Boolean = Random.nextBoolean()
while (condition) {
    println("Still true!")
```

```
        condition = Random.nextBoolean()

        thisMightChangeThings(condition)
    }
```

The good news is that while Kotlin initially allowed this sort of thing, it no longer does. So this is actually illegal Kotlin, and not something to worry about any longer.

> **NOTE** *Kotlin uses what is called "pass by value." Parameters are passed into functions as values, not "by reference" to the actual object or value in memory. That means that no function can change a value passed into it directly.*

But if you're writing code that is multithreaded, you could actually have a variable that is being accessed in multiple threads. That's a little complex to show, and well beyond what this chapter gets into. However, it's worth knowing: it is *possible* in a multithreaded environment for two different pieces of code to simultaneously be acting on a single variable. That case can create unpredictable changes to that variable, and if it is used in your `while` loop's condition, unpredictable evaluation of the condition.

> **NOTE** *This is a relatively brief, incredibly general overview of threading, and not one you should get too wrapped up in. It also doesn't mention coroutines, which are a big part of how Kotlin handles these situations.*
>
> *The main takeaway here is that you should know in rare edge cases, it's possible for your condition to get modified without you being aware. Consider yourself prepared now!*

Combine Control Structures for More Interesting Solutions

With an `if`, a `while`, and a `for`, you can start to do some interesting things. Remember, when your outer loop is a `while`, you are going to want something within your loop changing parts of the condition, so that the condition might change to false at some point.

Here's a bit of code that uses multiple loops to work out the prime numbers between `low` and `high`:

```
var low = 1
var high = 100

while (low < high) {
    var foundPrime = false

    for (i in 2 .. low/2) {
        if (low % i == 0) {
            foundPrime = true
            break
        }
    }
```

```
    if (!foundPrime) print("$low ")

    low++
}
```

> **NOTE** *This code is based on similar code from the Programiz site (www.pro-gramiz.com). That site is a great resource for Kotlin examples (as well as other languages). The sample also uses* break, *which as mentioned before, you'll learn shortly.*

DO . . . WHILE ALWAYS RUNS ONCE

A while loop always evaluates its condition before anything else occurs. So take this (trivial) example:

```
while (false) {
    println("Code that never runs!")
}
```

The println() line will never execute because the condition evaluates to false. But if you want your code to always execute at least once, you can switch to a do . . . while loop:

```
do {
    println("Gonna run at least once!")
} while (false)
```

Unlike while, the condition here is evaluated after execution of the code block. That is the only difference between while and do . . . while.

Every do . . . while Loop Can Be Written as a while Loop

You may have already figured this out, but it's an important point: every do . . . while loop can be rewritten as a while loop. And in every situation where you might want to use a do . . . while, you can also use a while loop. However, you may have to change the loop to make this conversion. So take this code:

```
condition = false
do {
    println("Gonna run at least once!")
} while (condition)
```

You cannot simply switch out do . . . while for while. So this won't mimic the previous code:

```
condition = false
while (condition) {
    // This line actually will NOT run now
    println("Gonna run at least once!")
}
```

In a do ... while, it doesn't matter what the initial value of `condition` is; the code will run once. In a while, condition's value does initially matter. So you'd need to fix that:

```
condition = true
while (condition) {
    // This line actually will run with condition initially set to true
    println("Gonna run at least once!")
    condition = false
}
```

Now you'll get the same output.

If Something Must Happen, Use do . . . while

It takes a little creative thinking to get your head around times that using do ... while makes more sense than while. As a rule of thumb, it's less about the code in the loop and more about the preexisting state of what that code operates upon.

For example, suppose you need to initialize a list with some data. First, here's a quick helper function to create random strings:

```
fun randomString(length: Int): String {
    val allowedChars = ('A'..'Z')
    return (1..length)
            .map { allowedChars.random() }
            .joinToString("")
}
```

> **NOTE** *There are some functions that you haven't seen here, like* map *and* join-ToString(), *that you don't need to worry too much about here. First, you can probably figure out what they do, and second, you'll learn more about* map *in particular in later chapters.*

Now, here's a do ... while loop that fills up an empty string with random strings:

```
var emptyList: MutableList<String> = mutableListOf()

do {
    emptyList.add(randomString(10))
} while (emptyList.size < 10)

println(emptyList)
```

There's really no reason to use a while and evaluate the condition first, because you *know* that the list is initially empty and has less than 10 items in it. This is a great case for do ... while, because you know you want the code to run at least once.

Now, this is actually still a little contrived, because you know the loop needs to run 10 times and could use a `for` loop. That would do what you want without needlessly checking the condition 9 times, when you know it's not really needed:

```
var anotherEmptyList: MutableList<String> = mutableListOf()

for (i in 1..10) {
    anotherEmptyList.add(randomString(10))
}

println(anotherEmptyList)
```

A better case is when you are operating on a variable and are *not* sure how many times that operation should occur. So perhaps you are handing off filling the array to a function that you don't have visibility into. Add a new utility function that creates a list with a random number of strings:

```
fun someStrings(): List<String> {
    var someStrings = mutableListOf<String>()
    for (i in 1..Random.nextInt(10))
        someStrings.add(randomString(10))
    return someStrings
}
```

Now, look at this modified version of the `do ... while` loop from earlier:

```
var emptyList: MutableList<String> = mutableListOf()

do {
    emptyList.addAll(someStrings())
} while (emptyList.size < 100)

println(emptyList)
```

The big difference here is that now it's not clear how many times the code will execute, but since `emptyList` starts as empty, it does need to execute at once. Those two things together make for a good `do ... while` use case:

➤ You know the state of the system such that the code in the loop *always* needs to run at least once

➤ You *don't* know how much or to what degree some part of the condition will be changed within the code block

Listing 7.3 updates Listing 7.2 to add all of these new example snippets.

LISTING 7.3: More code, more loops, more control structures

```
import kotlin.random.Random

fun main() {
    var theNumber = Random.nextInt(0, 1000)
```

```
print("Guess my number: ")
val guess = readLine()!!

/*
if (guess.toInt() == theNumber) {
    println("You got it!")
} else {
    println("Sorry, you missed it.")
}
 */

print("The number you guessed is ")
var hasRightNumber = if (guess.toInt() == theNumber) {
        println("right!")
        true
    } else {
        println("wrong.")
    false
    }

println(hasRightNumber)

var anotherNumber = Random.nextInt(0, 1000)

var even = false
when {
    (anotherNumber % 2 == 0) -> println("The number is even!")
    else -> println("The number is odd!")
}

when {
    (anotherNumber % 2 == 0) -> println("The number is divisible by 2!")
    (anotherNumber % 3 == 0) -> println("The number is divisible by 3!")
    (anotherNumber % 5 == 0) -> println("The number is divisible by 5!")
    (anotherNumber % 7 == 0) -> println("The number is divisible by 7!")
    (anotherNumber % 11 == 0) -> println("The number is divisible by 11!")
    else -> println("This thing is hardly divisible by anything!")
}

if (anotherNumber % 2 == 0) {
    println("The number is divisible by 2!")
} else if (anotherNumber % 3 == 0) {
    println("The number is divisible by 3!")
} else if (anotherNumber % 5 == 0) {
    println("The number is divisible by 5!")
} else if (anotherNumber % 7 == 0) {
    println("The number is divisible by 7!")
} else if (anotherNumber % 11 == 0) {
    println("The number is divisible by 11!")
} else {
    println("This thing is hardly divisible by anything!")
}

println("The actual number was $anotherNumber")
```

LISTING 7.3 *(continued)*

```kotlin
var foo: Long = 23

println(isPositiveInteger(foo))

val numbers = listOf(0, 1, 2, 3, 4)
for (item in numbers) {
    println(item)
}

val ar = intArrayOf(1, 2, 3, 4, 5, 6, 7, 8)
var i: Int
var x: Int

var counter: Int = 1
for (i in ar.indices) {
    x = ar[i]
    println("Item ${counter}: ${x} is at index ${i}")
    counter++
}

x = 1
while (x < 6) {
    println(x)
    x++
}

for (x in 1..5) {
    println(x)
}

outer_loop@ for (i in 1..10) {
    var low = 1
    var high = 100

    while (low < high) {
        var foundPrime = false

        prime_loop@ for (i in 2..low / 2) {
            if (low % i == 0) {
                foundPrime = true
                break@outer_loop
            }
        }

        if (!foundPrime) print("$low ")

        low++
    }
    println()
}
```

```
var condition: Boolean = Random.nextBoolean()
while (condition) {
    println("Still true!")
    condition = Random.nextBoolean()

    // thisMightChangeThings(condition)
}
println("Not true any more!")

while (false) {
    println("Code that never runs!")
}

condition = false
do {
    println("Gonna run at least once!")
} while (condition)

condition = false
while (condition) {
    // This line actually will NOT run now
    println("Gonna run at least once!")
}

condition = true
while (condition) {
    // This line actually will NOT run now
    println("Gonna run at least once!")
    condition = false
}

var emptyList: MutableList<String> = mutableListOf()

do {
    emptyList.addAll(someStrings())
} while (emptyList.size < 100)

println(emptyList)
```

do . . . while Can Be a Performance Consideration

Now that you've seen while and do ... while, and realize that you really *can* use while for everything, it raises the question: why *ever* use do ... while? Certainly, even in the examples you've seen and written, the difference is small and arguably negligible.

That's actually true. Given that, you have only really two good reasons to consider do ... while over while:

➤ You want to make the intention of your code as clear as possible, and you know that the code in your block will run at least once (and it's potentially useful to make that clear to others who read, review, or maintain your code).

➤ The negligible difference between while and do ... while matters, and you anticipate that difference adding up over time.

You may not often find yourself coding in one of these two contexts, so using `while` as your standard structure is absolutely OK. However, if you *do* find yourself in a place where performance is critical, it tends to matter *a lot*. In other words, that small percentage of the time that the difference between `do ... while` and `while` executing at least once versus maybe not at all tends to really be a game changer.

GET OUT OF A LOOP IMMEDIATELY WITH BREAK

So far, you've seen a number of control structures. Each allows you to either choose from one of a few different paths or repeat a portion of a path over and over. These various structures and iterations can be simple or complex, and you can get pretty fancy with conditions and iterators if you need to.

The last two sections in this chapter focus on a bit more of a heavy-handed approach to controlling the flow of your code. Using `break` (covered in this section) and `continue` (covered in the next) are great tools, but used less often than `if` and `for` and `while`. They are blunt instruments more often than not, ripping you out of the sequence of code being executed, potentially skipping over lines or jumping from one bit of code to another a bit more abruptly than you've seen so far.

Break Skips What's Left in a Loop

Suppose you have a loop like this:

```
var low = 1
var high = 100

while (low < high) {
    var foundPrime = false

    for (i in 2 .. low/2) {
        if (low % i == 0) {
            foundPrime = true
            break
        }
    }

    if (!foundPrime) print("$low ")

    low++
}
```

You saw this code earlier in an example of using multiple control flow structures in a single piece of code. This code hunts through a range of numbers (between low and high) and sees if each number it examines is prime. Once it finds a prime, the loop needs to stop executing.

This is important, because finding a prime is effectively a *second* exit condition for the loop. The first exit condition is the iteration from 2 to `low/2`. Of course, `for` loops don't offer two exit conditions, which is where `break` comes in.

Using that keyword with an `if`, there's this extra exit condition:

```
if (low % i == 0) {
    foundPrime = true
    break
}
```

If this condition evaluates to true, then `foundPrime` is updated, and the `break` keyword exits *the nearest enclosing loop*. That's the `for` loop here:

```
for (i in 2 .. low/2) {
    if (low % i == 0) {
        foundPrime = true
        break
    }
}
```

You Can Use a Label with break

Sometimes, as with most control structures, it can be hard to know exactly where program execution goes. This is particularly true with `break`, as control does not continue to the next line, or move to a specific block of code further down in your code.

To aid in this, you can use a label. A label in Kotlin is a single word that ends with @. That word that is ignored by the Kotlin compiler in terms of syntax. But it can be used by other pieces of your code.

Here's that same for loop as earlier, with an added label:

```
prime_loop@ for (i in 2 .. low/2) {
    if (low % i == 0) {
        foundPrime = true
        break
    }
}
```

You can now use that label in your `break` statement. You use `break` as you normally would, then add the @ sign, and then the name of the label:

```
prime_loop@ for (i in 2 .. low/2) {
    if (low % i == 0) {
        foundPrime = true
        break@prime_loop
    }
}
```

> **WARNING** *The use of @ in Kotlin syntax with labels and* break *can be a little confusing at first. The @ character goes* after *the loop name, but* before *the loop name when used in a* break. *This just takes some time before it becomes natural.*

In this particular example, the label isn't needed; break already was jumping out of that for loop, because it's the nearest currently executing for loop. In one sense, the label does add clarity: it indicates the loop that break is getting your code out of. So control would continue with the first line *after* the for loop marked with the matching label.

Now, that's still not particularly clear, so using a label "just because" won't always help your code's legibility that much.

Here's an example with two loops, each with a distinct label:

```
outer_loop@ for (j in 1..10) {
    var low = 1
    var high = 100

    while (low < high) {
        var foundPrime = false

        prime_loop@ for (i in 2..low / 2) {
            if (low % i == 0) {
                foundPrime = true
                break@outer_loop
            }
        }

        if (!foundPrime) print("$low ")

        low++
    }
    println()
}
```

In this example, when a prime is found, break uses the outer_loop label to jump out of the for loop running from 1 to 10—which is *not* the nearest executing loop.

> **NOTE** *Full disclosure: finding a practical example for using a labeled break is actually pretty tough. I couldn't think of a single case where I've used something like this in nearly 20 years of programming, and even some significant time on Google largely turned up either contrived or downright silly examples. So know that you can do this, but realize you might never actually use it!*

GO TO THE NEXT ITERATION IMMEDIATELY WITH CONTINUE

break is a relatively extreme flow control because it immediately (and in some cases, abruptly) jumps out of the current or labeled loop. continue, on the other hand, moves to the next *iteration* of the current loop. So it stops the current iteration—ignoring any remaining code in that loop—and continues with the next iteration.

Here's a simple loop that uses `continue`:

```
for (i in 1..20) {
    if (i % 2 == 0) continue
    println("${i} is odd")
}
```

This loop has an `if`, and if the condition is true, the `continue` goes to the next iteration. That means for even numbers, this statement will be skipped:

```
println("${i} is odd")
```

This is the significant advantage of `continue`: it leaves your loop intact. You write the loop body just as you normally would, knowing that any code after a `continue` is simply ignored in that iteration.

Continue Works with Labels as Well

Just as you can use a `break` with a labeled loop, you can do the same with `continue`:

```
measure@ for (measure in 1..10) {
    print("Measure ${measure}: ")
    for (beat in 1..4) {
        // Lay out on beats 3 and 4 of measures 5 and 10
        if ((measure == 5 || measure == 10) && (beat in 3..4)) {
            println(".....")
            continue@measure
        }


        // only "play" on the 2 and 4
        if (beat % 2 == 0)
            print("snare ")
        else
            print("${beat} ")
    }
    println()
}
```

This loop has a few control structures and uses a `continue` to jump out of the outer loop (the `for` using `measure` as a counter) when certain conditions are met.

If versus continue: Mostly Style over Substance

Look carefully again at this code:

```
for (i in 1..20) {
    if (i % 2 == 0) continue
    println("${i} is odd")
}
```

You could just as easily rewrite this code using an `if` without `continue`:

```
for (i in 1..20) {
    if (i % 2 != 0) println("${i} is odd")
}
```

The condition for the `if` has to be negated (using `!=` instead of `==`), but then the `if` statement conditionally executes the `println`, and you get the same result.

In fact, you can almost always use an `if` instead of a `continue` if you like. Suppose a loop looks like this:

```
for (CONDITION) {
    // Code that always runs
    if (CONDITION) {
        continue
    }
    // Code that only runs if the above CONDITION was false
}
```

This loop can be converted to use `if` and *not* use `continue` like this:

```
for (CONDITION) {
    // Code that always runs
    if (!CONDITION) {
        // Code that only runs if the above CONDITION was true
    }
}
```

So then why use one over the other? Well, it's really a matter of preference. You're always going to have some sort of condition associated with `continue`, or you'd never complete your loop body. That means you can use the opposite of that same condition to execute the rest of the loop body. Either way, the effect is the same.

The primary advantage to using `continue` is that it's very clear what's going on, and you don't have to put a lot of code into an `if` block. If you have 20 or 30 lines of code that only executes in certain situations, and especially if that code has control structures of its own, `continue` can really clean up your code.

On the other hand, if you are only skipping a line or two with `continue`, it actually is easier to use an `if` without `continue`. This will make your code simpler to read.

RETURN RETURNS

It's worth mentioning `return` briefly, even though you've been using `return` for quite a bit already. Put simply, `return` returns. More technically, `return` exits out of the nearest function body.

For the most part, you've seen `return` used at the end of a function, like this:

```
fun someStrings(): List<String> {
    var someStrings = mutableListOf<String>()
    for (i in 1..Random.nextInt(10))
        someStrings.add(randomString(10))
    return someStrings
}
```

But you can put a return anywhere. Here's another simple utility function to find a string in a supplied list that starts with a particular character:

```
fun findFirstStringStartingWith(starts: Char, strings: List<String>): String {
    for (str in strings) {
        if (str.startsWith(starts)) return str
    }
    return ""
}
```

return is used here to jump out of this code as soon as a matching string is found. That could result in a big change in how this code executes. Suppose you had a list of 1,000 strings, and the second string started with the desired character. By using return, you skip another 998 iterations through this loop.

So what's next? You've just blown through over 30 pages of new control structures and ways to adjust the program flow. With that in hand, it's time to go back to class and dig even deeper into what Kotlin provides.

8

Data Classes

WHAT'S IN THIS CHAPTER?

➤ Modeling functionality with classes

➤ Data classes for efficiency (and more)

➤ Destructuring data from data classes

CLASSES IN THE REAL WORLD ARE VARIED BUT WELL EXPLORED

If you learned to program in a language like Java, and if you've been around programming for more than 5 or 6 years, you have an important perspective in programming: you have quite literally seen the evolution of programming languages—and especially object-oriented programming languages. This chapter in many ways is the result of decades of programmers working with classes and objects.

Many Classes Share Common Characteristics

Initially, classes were as blank a slate as possible. You often see classes just like the ones you've been building. Person, as a great example, is shown again in Listing 8.1. This is a pretty typical class.

LISTING 8.1: Revisiting the Person class

```
package org.wiley.kotlin.person

open class Person(_firstName: String, _lastName: String,
                  _height: Double = 0.0, _age: Int = 0) {
```

continues

LISTING 8.1 *(continued)*

```
var firstName: String = _firstName
var lastName: String = _lastName
var height: Double = _height
var age: Int = _age

var partner: Person? = null

constructor(_firstName: String, _lastName: String,
            _partner: Person) :
            this(_firstName, _lastName) {
partner = _partner
    }

    fun fullName(): String {
        return "$firstName $lastName"
    }

    fun hasPartner(): Boolean {
        return (partner != null)
    }

    override fun toString(): String {
        return fullName()
    }

    override fun hashCode(): Int {
        return (firstName.hashCode() * 28) + (lastName.hashCode() * 31)
    }

    override fun equals(other: Any?): Boolean {
        if (other is Person) {
            return (firstName == other.firstName) &&
                   (lastName == other.lastName)
        } else {
            return false
        }
    }
}
```

For the most part, this is as typical a class as you might hope to find. It has the following very common characteristics:

➤ The class implicitly inherits directly from `Any`.

➤ The class is largely a set of accessors (getters) and mutators (setters) on a number of key properties.

➤ Those properties can be set in the constructor and also changed later in the class instance's existence.

➤ The class overrides `hashCode()` and `equals(x)` to ensure valid usage of both methods—using the properties of the class.

Common Characteristics Result in Common Usage

Now, because this is such a common set of characteristics, programmers have been creating classes like these and using them for nearly as long as object-oriented programming has existed. That's created a sort of shared body of knowledge around classes like this; a set of typical use cases that have been tested through time and experience.

> **NOTE** *It's worth saying here that in addition to simply reading other program-mers' code, you should try to understand why other programmers write their code the way that they do. It's great to program as much as you can; you'll inevi-tably learn your own lessons. It's even better to learn from others who have also been programming for a long time, and build a common set of what are typically called* best practices *as to how you code and use Kotlin or any other language.*

The end result is that modern languages (like Kotlin) and modern versions of older languages (like Java) are starting to take those commonalities into account, and introduce them into the language itself. In the case of this set of characteristics, that means a new type of class: the data class.

A DATA CLASS TAKES THE WORK OUT OF A CLASS FOCUSED ON DATA

That heading sounds a little circular. A better approach is to simply look at a new take on a class definition for yourself. Listing 8.2 shows a new *data class* in Kotlin.

LISTING 8.2: A new User class as a data class

```
package org.wiley.kotlin.user

data class User(val email: String, val firstName: String, val lastName: String)
```

That's a remarkably terse class. It also seems to be missing most of the things you have grown to associate with classes: curly braces, code in the constructor, any secondary constructors, a useful `toString()` method, a functional `hashCode()` method . . . in fact, pretty much everything that `Per-son` has in Listing 8.1, `User` does *not* have in Listing 8.2.

But it will absolutely compile. Try it!

Data Classes Handle the Basics of Data for You

To get a sense of what you can do with `User`—all with just a few lines of code—create a test class. You can call it `DataClassApp` if you want to follow along with the example code. Listing 8.3 shows this class's starting point.

LISTING 8.3: Testing out the User data class

```kotlin
import org.wiley.kotlin.user.User

fun main() {
    val robert = User("powell@rockwallblackbelt.com", "Robert", "Powell")

    println(robert)
}
```

Run this class and you'll get the following output:

```
User(email=powell@rockwallblackbelt.com, firstName=Robert, lastName=Powell)
```

There's definitely a lot going on under the hood. But there's quite a bit more you can do "for free." Make these additions to your test class:

```kotlin
fun main() {
    val robert = User("powell@rockwallblackbelt.com", "Robert", "Powell")

    robert.email = "rpowell@rockwallblackbelt.com"
    robert.firstName = "Bob"

    println(robert)
}
```

Now recompile, and you'll get these two errors:

```
Error:(6, 5) Kotlin: Val cannot be reassigned
Error:(7, 5) Kotlin: Val cannot be reassigned
```

That makes sense; look again at the declaration of `User`, which did indeed use `val`:

```kotlin
data class User(val email: String, val firstName: String, val lastName: String)
```

If you want these properties to be changeable, you need to use `var`:

```kotlin
data class User(var email: String, var firstName: String, var lastName: String)
```

Now recompile and the errors will go away. You'll also notice that when the `robert` instance is printed, the updated values for `email` and `firstName` are used:

```
User(email=rpowell@rockwallblackbelt.com, firstName=Bob, lastName=Powell)
```

So without writing anything but a class declaration, you've already gotten a lot of basic code:

➤   Any properties defined with `val` get accessor (getter) methods

➤   Any properties defined with `var` get mutator (setter) and accessor methods

➤   `toString()` works immediately and uses the property values of the class

The Basics of Data Includes hashCode() and equals(x)

Remember that for quite a while now, `hashCode()` and `equals(x)` have become an important part of how you think about classes. But also remember that you don't get meaningful versions of these methods when you create an empty class—at least until now.

To drive this point home, Listing 8.4 shows a simple version of `User`—called `SimpleUser`—that is *not* a data class.

LISTING 8.4: A non–data class version of User

```
package org.wiley.kotlin.user

class SimpleUser(_email: String, _firstName: String, _lastName: String) {
    var email: String = _email
    var firstName: String = _firstName
    var lastName: String = _lastName
}
```

This class is, at least on the surface, functionally identical to User. Listing 8.5 is an updated version of the test class to compare how User (a data class) compares to SimpleUser (which isn't a data class).

LISTING 8.5: Comparing what you get "for free" between User and SimpleUser

```
import org.wiley.kotlin.user.SimpleUser
import org.wiley.kotlin.user.User

fun main() {
    val robert = User("powell@rockwallblackbelt.com", "Robert", "Powell")
    val simpleRobert = SimpleUser("powell@rockwallblackbelt.com",
"Robert", "Powell")

    robert.email = "rpowell@rockwallblackbelt.com"
    robert.firstName = "Bob"

    simpleRobert.email = "rpowell@rockwallblackbelt.com"
    simpleRobert.firstName = "Bob"

    println(robert)
    println(simpleRobert)
}
```

The output is immediately and notably different:

```
User(email=rpowell@rockwallblackbelt.com, firstName=Bob, lastName=Powell)
org.wiley.kotlin.user.SimpleUser@58ceff1
```

That's the default version of toString() for SimpleUser, and of course User as a data class as a nicer version.

But there's more! Data classes also give you a working and useful equals(x) method based on properties of the data class. Try this:

```
fun main() {
    val robert = User("powell@rockwallblackbelt.com", "Robert", "Powell")
    val simpleRobert = SimpleUser("powell@rockwallblackbelt.com",
"Robert", "Powell")

  .  robert.email = "rpowell@rockwallblackbelt.com"
    robert.firstName = "Bob"
    val robert2 = User(robert.email, robert.firstName, robert.lastName)
```

```
        simpleRobert.email = "rpowell@rockwallblackbelt.com"
        simpleRobert.firstName = "Bob"
        val simpleRobert2 = SimpleUser(simpleRobert.email, simpleRobert.firstName,
    simpleRobert.lastName)

        println(robert)
        if (robert.equals(robert2)) {
            println("robert and robert2 are equal")
        } else {
            println("robert and robert2 are not equal")
        }

        println(simpleRobert)
        if (simpleRobert.equals(simpleRobert2)) {
            println("simpleRobert and simpleRobert2 are equal")
        } else {
            println("simpleRobert and simpleRobert2 are not equal")
        }
    }
```

The output might surprise you:

```
User(email=rpowell@rockwallblackbelt.com, firstName=Bob, lastName=Powell)
robert and robert2 are equal
org.wiley.kotlin.user.SimpleUser@58ceff1
simpleRobert and simpleRobert2 are not equal
```

As expected, `simpleRobert` and `simpleRobert2` aren't equal. Remember, object equality is effectively an in-memory object comparison by default, and since `simpleRobert` and `simpleRobert2` aren't the same actual object, they're not considered equal.

But `robert` and `robert2`, both data class instances, *do* report as equal. Their properties are identical and because of the "for free" implementation you get of `equals(x)`, they report as being equal.

As you should also expect, where there's a modified `equals(x)`, there's an accompanying `hashCode()`. It's not very interesting to see these (you're welcome to do this as an exercise on your own), but data classes also provide useful `hashCode()` method implementations.

DESTRUCTURING DATA THROUGH DECLARATIONS

There's another interesting bit of functionality you get with data classes, but it's going to require you to take a step back and learn some new concepts first. The idea here is that Kotlin allows you to do something called destructuring declarations, and it mostly comes up when you have objects that are primarily holders of data—like a data class.

In this model, you can actually treat an entire class as a collection of data, without regard to the methods or functionality of the class.

Grab the Property Values from a Class Instance

Add this code to your test class:

```
val (email, firstName, lastName) = robert
```

> **WARNING** *This code will work as shown as long as you've been following along with the code samples in this chapter so far, and as long as this code appears after the* robert *instance of* User *has been instantiated. If you've not been following along, just create a new instance of a data class, make sure the properties of the class instance are set, and then you can destructure it as shown in the preceding code.*

This looks a bit odd, but if you stare at it a bit, it will start to make some sense. Three variables are created: email, firstName, and lastName. Then, they are all assigned values *at once* through a *destructuring declaration*. This declaration takes the property values from robert and assigns each to the three variables.

Here's the output:

```
Email is rpowell@rockwallblackbelt.com
First name is Bob
Last name is Powell
```

This is a case where Kotlin again is helping you simply grab data in a way that's convenient and quick.

Destructuring Declarations Aren't Particularly Clever

Now before you get too excited, try this bit of code, which looks a lot like the previous code, but is in fact quite different:

```
val(firstName, lastName, email) = robert
println("Email is ${email}")
println("First name is ${firstName}")
println("Last name is ${lastName}")
```

The output here is not what you'd hope for:

```
Email is Powell
First name is rpowell@rockwallblackbelt.com
Last name is Bob
```

So what's going on? Well, Kotlin has no idea what a variable named email means; it's just a sequence of letters. So when you have a destructuring declaration, Kotlin is *not* matching up object property names to variable names. You could just as easily have code like this:

```
val(a, b, c) = robert
```

This code works, and a will have email, b will have firstName, and c will have lastName. Kotlin is simply taking the first property in User and assigning it to the first variable in the destructuring declaration, taking the second property and assigning it to the second variable, and so on. It's a bit of a dumb operation, albeit a *useful* dumb operation.

In fact, if you changed your data class, you'd also change what gets assigned. So suppose you have this as your opening line of `User`:

```
data class User(var firstName: String, var lastName: String, var email: String)
```

Now assume that same bit of code:

```
val(a, b, c) = robert
```

Now a has the value of `firstName`, b has `lastName`, and c has `email`.

> **NOTE** *If you want to keep following along—and that's highly encouraged—make sure that you return* email *to the first property of* User, firstName *to the second, and* lastName *to the third. That will keep your output consistent with the rest of this chapter.*

Kotlin Is Using componentN() Methods to Make Declarations Work

Before you go getting too excited, you should realize that using destructuring declarations is not automatic; in other words, it's not something that any class supports. In fact, go back to your `SimpleUser` class (shown back in Listing 8.4). Try this code in your test class:

```
val(email, firstName, lastName) = simpleRobert
println("Email is ${email}")
println("First name is ${firstName}")
println("Last name is ${lastName}")
```

> **NOTE** *Make sure you change the first line to use* simpleRobert *(or any other instance of* SimpleUser*) instead of* robert, *which was an instance of* User, *your data class.*

The compiler won't like this, and tells you about it:

```
Error:(30, 39) Kotlin: Destructuring declaration initializer of type SimpleUser
must have a 'component1()' function
Error:(30, 39) Kotlin: Destructuring declaration initializer of type SimpleUser
must have a 'component2()' function
Error:(30, 39) Kotlin: Destructuring declaration initializer of type SimpleUser
must have a 'component3()' function
```

So what's going on here? Well, the error actually is giving you a good idea: Kotlin uses what is often called a `componentN()` method to support destructuring declarations. N just represents a number. So to support a destructuring declaration with three properties, the class would need a `component1()` method, a `component2()` method, and a `component3()` method.

In fact, look again at a destructuring declaration that works (using an instance of User rather than SimpleUser):

```
val(email, firstName, lastName) = robert
```

This code could be rewritten like this:

```
val email = robert.component1()
val firstName = robert.component2()
val lastName = robert.component3()
```

In fact, this compiles! The User class has actual methods for all of its properties; that's often stated as "data classes provide componentN() methods."

You Can Add componentN() Methods to Any Class

Data classes give you these methods for free, but that doesn't mean you can't add them to any other class. Listing 8.6 is an expanded version of SimpleUser that does just that.

LISTING 8.6: Adding componentN() methods to SimpleUser

```
package org.wiley.kotlin.user

class SimpleUser(_email: String, _firstName: String, _lastName: String) {
    var email: String = _email
    var firstName: String = _firstName
    var lastName: String = _lastName

    operator fun component1(): String {
        return email
    }

    operator fun component2(): String {
        return firstName
    }

    operator fun component3(): String {
        return lastName
    }
}
```

NOTE *You'll need to use the* operator *keyword for* component() *functions.*

Now you can take code that didn't work earlier and use it successfully:

```
val(email, firstName, lastName) = simpleRobert
```

If You Can Use a Data Class, You Should

This really highlights the value of data classes. While it wasn't a tremendous amount of work to add a `component1()`, `component2()`, and `component3()` method to `SimpleUser`, the upkeep is significant. Just think of how often these methods might need to change:

➤ A new property is added to a class and therefore a new `componentN()` method is needed.

➤ The order of properties changes and therefore the body of the methods needs to change to return the correct property.

➤ A property is removed and therefore the accompanying component method needs to also be removed.

Compare this to a data class. If you're using a data class and *any* of the preceding situations occur, you have to do . . . nothing. Just recompile, and Kotlin takes care of the rest.

YOU CAN "COPY" AN OBJECT OR MAKE A COPY OF AN OBJECT

What a mouthful of a heading! But it's true: you have two different ways to create a copy of an object instance, and data classes are really helpful in this regard.

Using = Doesn't Actually Make a Copy

First, you can copy through a simple = assignment. This is actually not creating a copy (and why the heading for this section reads "copy" with quotation marks). Here's how you'd typically see that:

```
val bob = robert
```

This looks straightforward but has some issues. It's actually not creating a copy of anything. Instead, you now have two variables: `robert` and `bob`. Both point at the same bit of memory, and both point to the same actual instance of `User`. That means that changes to one affect the other.

Try out the code shown in Listing 8.7.

LISTING 8.7: Using = instead of copy()

```
// Create another variable pointed at the robert instance
val bob = robert

// Change things and see the effect
bob.firstName = "Percy"
println("Bob's first name: ${bob.firstName}")
println("Robert's first name: ${robert.firstName}")
robert.firstName = "Bob"
println("Bob's first name: ${bob.firstName}")
println("Robert's first name: ${robert.firstName}")
```

Here's the output:

```
Bob's first name: Percy
Robert's first name: Percy
Bob's first name: Bob
Robert's first name: Bob
```

The most important thing here is to realize that the actual object instance lives in memory and that the `robert` variable just references that memory. And when you use = to assign `robert` to a new variable—`bob`—you're just assigning that reference to `bob`.

When you change the `firstName` property of `bob`, you're changing it in what `bob` actually points at—which happens to be what `robert` points at, too. So changes are reflected in both variables, and in fact, in *any* variables that reference that same instance in memory.

If You Want a Real Copy, Use copy()

If you want an *actual* copy, you can't use =. You have to use the `copy()` method, which is another thing you get for free with data classes.

`copy()` works just like you'd think. You get a new variable with a new instance, but that instance starts with all the values of the old instance. Here's how that looks:

```
// Create a variable that is a copy of the robert instance
val differentBob = robert.copy()
```

> **NOTE** *Non-data classes don't get a* copy() *implementation that's useful by default, so make sure you're using* copy() *on a data class instance.*

What's actually happening here is that there are now *two* object instances in play: the original one, which `robert` points to, and a new one—with its own memory space—which `bob` points to.

You can see this in action with some more test code, as shown in Listing 8.8.

LISTING 8.8: Moving from = to copy()

```
// Create a variable that is a copy of the robert instance
val differentBob = robert.copy()

// Change the original instance
robert.firstName = "Robert"
robert.email = "powell@rbba.com"

println("Bob's email is ${bob.email}")
println("Bob's first name is ${bob.firstName}")
println("Bob's last name is ${bob.lastName}")

println("Different Bob's email is ${differentBob.email}")
println("Different Bob's first name is ${differentBob.firstName}")
println("Different Bob's last name is ${differentBob.lastName}")
```

The output shows that these two instances are now completely independent:

```
Bob's email is powell@rbba.com
Bob's first name is Robert
Bob's last name is Powell
Different Bob's email is rpowell@rockwallblackbelt.com
Different Bob's first name is Bob
Different Bob's last name is Powell
```

In fact, after the initial `copy()`, there is no connection at all between `bob` and `robert`.

DATA CLASSES REQUIRE SEVERAL THINGS FROM YOU

Clearly, data classes offer a lot of advantages. Additionally, you'll find that a lot of your classes will almost seamlessly transition to data classes. Just add the `data` keyword before `class` in your class declaration, and you're ready to go:

```
data class User(var email: String, var firstName: String, var lastName: String)
```

But there are some requirements, albeit ones that are typically easy to meet.

Data Classes Require Parameters and val or var

In the primary constructor for your data class, you must have at least one parameter. That means that this is not valid Kotlin:

```
data class User
```

You need something in that primary constructor, a named parameter like `email` or `firstName`:

```
data class User(var email: String, var firstName: String, var lastName: String)
```

This should make perfect sense, though. A data class is about data, and if your class doesn't actually represent any data—which is fed to the class through parameters in the primary constructor—then the whole idea breaks down.

Additionally, the parameters to the class's primary constructor must be either `var` or `val`. This is also sensible given that a data class really is just an object designed to easily store property values. `var` tells the data class that the property can be both set and get (mutable), and `val` tells the data class that the property is read-only (immutable).

As a reminder, the situation where this would *not* be the case would be when a primary constructor is taking in values that are used to create a new value. For example, suppose you created a class that took in a set of values and only stored the minimum, and maximum, and also calculated the average:

```
class Calculator(values: Set)
```

This class would then presumably do some calculations using the values in the set in its constructor. Since the actual values aren't being stored (in this rather silly hypothetical), the `values` property isn't preceded by either `val` or `var`.

This would not work as a data class, though. To make that work, values would have to be stored:

```
data class Calculator(val values: Set)
```

> **NOTE** *It should be said outright that this example is a bit silly. There's almost no reason why a class that uses values like this would not store those values as a property. In fact, it's arguable that a class like* Calculator *that uses a set of values is poorly designed if it does* not *store the input values.*

Data Classes Cannot Be Abstract, Open, Sealed, or Inner

Another requirement of data classes is that they cannot be abstract, open, sealed, or inner. You've seen open classes already, but not gotten into sealed, inner, or abstract classes. Don't worry, though: coverage of these is coming up in Chapter 9.

Most of this boils down to this simple idea: data classes can inherit from a non-data class, but cannot themselves be inherited from. If you create a User data class, you won't be able to inherit from it and create a SuperUser class. That's the commonality between abstract, open, and sealed and inner classes that's relevant here: they are set up to easily be inherited from.

The problem, of course, is that data classes are almost entirely composed of compiler-generated data methods and properties that are not explicitly declared. But when you inherit from a class, those are the very things you'll typically override or extend. There's simply no consistent way for a subclass to modify compiler-generated code without knowing more about that code than Kotlin makes explicitly apparent.

It is still legal for a data class to inherit from a superclass, though. This also should be pretty intuitive, as by now you know that *all* Kotlin classes implicitly inherit from Any.

> **WARNING** *With the exception of the implicit inheritance from* Any, *you should try to avoid creating a data class that inherits from another custom class, or in fact any class outside of* Any. *A lot of tricky logic can result from the collision of the compiler-generated code in a data class and normal inheritance.*
>
> *There's also growing discussion and movement toward disallowing inheritance in data classes. While this is still legal as of this writing, it may not be in future versions of Kotlin, so better to get ahead of things now and just allow data classes to stand on their own.*

DATA CLASSES ADD SPECIAL BEHAVIOR TO GENERATED CODE

In addition to the restrictions on them, data classes add a number of wrinkles to normal rules of inheritance and some of the special methods you've already learned about, such as toString() and hashCode().

You Can Override Compiler-Generated Versions of Many Standard Methods

Data classes will automatically give you meaningful and usable versions of equals(), hashCode(), and toString(). You saw that earlier in Listings 8.2, 8.3, 8.4, and 8.5. But that doesn't prevent you from creating your own versions of these functions.

All you have to do is define your own version of these functions, and Kotlin will simply not generate a version of that function. Listing 8.9 adds a custom toString() to User.

Listing 8.9: A data class with an overridden toString()

```
package org.wiley.kotlin.user

data class User(var email: String, var firstName: String, var lastName: String) {

    override fun toString(): String {
        return "$firstName $lastName with an email of $email"
    }
}
```

If you execute this function, such as with println(robert) (assuming of course that robert is an instance of User), you'll get something like this:

```
Bob Powell with an email of rpowell@rockwallblackbelt.com
```

Kotlin uses your version of toString() and doesn't generate one of its own.

The same is true for equals() and hashCode(). Override them yourself, and Kotlin will use your versions. However, the same best practices as mentioned in several earlier chapters apply: if you override equals(), you *must* override hashCode() with an appropriate corresponding version of code.

> **NOTE** *Be careful when overriding these compiler-generated methods. You give up a lot of the value of a data class when you write code for the things that Kotlin is generating for you.*

While you can override equals(), hashCode(), and toString(), you cannot override copy() or componentN() functions. These must remain compiler-generated.

Supertype Class Functions Take Precedence

In general, if you create a data class that inherits from a superclass, any of the superclass functions marked as final result in the data class not generating those methods. So if a class that a data class inherits from had a final version of toString(), that version would be used in the data class as well.

> **NOTE** *You've not seen the* final *keyword, but it does just what it sounds like it does: marks a chunk of code as "final," and disallows subclasses to override or replace that code.*

If a superclass has componentN() functions, things get a little trickier. As long as the componentN() functions are marked as open and return compatible types, the data class doesn't generate code for those functions. But if they are not open, or are incompatible in return type, then Kotlin will throw an error, because the data class won't be able to generate legal code that works with the superclass.

If the superclass has a copy() function with the signature the data class would use, then things break altogether. Kotlin will throw an error and you'll have to either remove the inheritance or remove or change the superclass's copy() function.

> **WARNING** *Take note that if you follow the earlier advice of avoiding creating a data class that inherits from any class other than* Any, *these last several sections aren't things you need to worry about. That's the best approach, and the simplest one as well: keep your data classes as simple as possible, and override as little as possible.*

Data Classes Only Generate Code for Constructor Parameters

You've seen in several cases now that the constructor isn't the only way to get data into a class. You can also declare properties inside the class body. These are usually derived from constructor parameters. Listing 8.10 is an older version of the Person class that had code with a property declared inside the class.

LISTING 8.10: Person with extra properties

```
package org.wiley.kotlin.person

class Person(var firstName: String, var lastName: String) {
    var fullName: String

    // Set the full name when creating an instance
    init {
        fullName = "$firstName $lastName"
    }

    override fun toString(): String {
        return fullName
    }
}
```

Here, the `fullName` property is assigned a value based on two parameters to the constructor, `firstName` and `lastName`. You could turn this into a data class if you wanted, as shown in Listing 8.11.

LISTING 8.11: Person with extra properties, converted to a data class

```
package org.wiley.kotlin.person

data class Person(var firstName: String, var lastName: String) {
    var fullName: String

    // Set the full name when creating an instance
    init {
        fullName = "$firstName $lastName"
    }

    override fun toString(): String {
        return fullName
    }
}
```

Kotlin will generate functions to access and mutate `firstName` and `lastName`. It will also use the overridden version of `toString()` instead of generating one. What Kotlin will *not* do is generate functions to interact with `fullName`. So this code would work:

```
val brian = Person("Brian", "Truesby")
brian.lastName = "Tannerton"
```

This code would also work, but *not* because `Person` is now a data class:

```
val brian = Person("Brian", "Truesby")
println(brian.fullName)
```

This bit of code works because Kotlin always sets up access to properties that you declare in this manner. However, `fullName` would *not* be used in generated methods, like `equals()`, `hashCode()`, and `toString()`.

To better see this, remove the overridden version of `toString()`, as shown in Listing 8.12.

LISTING 8.12: Person as a data class with a declared property

```
package org.wiley.kotlin.person

data class Person(var firstName: String, var lastName: String) {
    var fullName: String

    // Set the full name when creating an instance
    init {
        fullName = "$firstName $lastName"
    }
}
```

> **NOTE** *Assuming you're following along, the only change here from Listing 8.10 is a removal of code.*

Now, create a new instance of `Person` and display the `toString()` message:

```
val brian = Person("Brian", "Truesby")
println(brian)
```

The output will look like this:

```
Person(firstName=Brian, lastName=Truesby)
```

For comparison, now update the data class version of `Person` to accept `fullName` as a parameter to the constructor, as shown in Listing 8.13.

LISTING 8.13: Person as a data class with all properties accepted in the constructor

```
package org.wiley.kotlin.person

data class Person(var firstName: String, var lastName: String, var
fullName: String) {
    // var fullName: String

}
```

Perform the same instance creation and printing as you did before:

```
val brian = Person("Brian", "Truesby", "Brian T. Truesby")
println(brian)
```

The output of `toString()` now adds in the additional parameter:

```
Person(firstName=Brian, lastName=Truesby, fullName=Brian T. Truesby)
```

That's because with `fullName` as a parameter to the constructor, it gets included automatically when the compiler generates `toString()` code. The same is true for `equals()` and `hashCode()`.

In this case, that's probably OK. Since `fullName` is a completely derived property, just accessing it is enough. It doesn't need to be part of `equals()`, `hashCode()`, `toString()`, and the like.

Only Constructor Parameters Are Used in equals()

There's a more subtle aspect of generated code only using properties that are in the primary constructor: the generated `equals()` method only compares properties passed into that constructor.

Listing 8.14 is another simple variation on the `Person` class. In this version, `Person` is a data class with two parameters passed into the constructor and an additional parameter, `numChildren`. You can imagine this using internal properties to derive this value.

LISTING 8.14: Person as a data class with a single internal property

```
package org.wiley.kotlin.person

data class Person(var firstName: String, var lastName: String) {
    var numChildren: Int = 0
}
```

Kotlin will generate all functions based on the `firstName` and `lastName` property, including `equals()`. These functions will *not* use `numChildren` at all in those functions.

Now consider this code:

```
val jason = Person("Jason", "Smith")
jason.numChildren = 2

val anotherJason = Person("Jason", "Smith")
anotherJason.numChildren = 1
```

Two instances of `Person` are created, with the same name, but with a different number of children. These instances are *not* equal if you consider the number of children they each have. However, try comparing them with `equals()`:

```
val jason = Person("Jason", "Smith")
jason.numChildren = 2

val anotherJason = Person("Jason", "Smith")
anotherJason.numChildren = 1

println(jason.equals(anotherJason))
```

The output here is simple:

```
true
```

The generated `equals()` considers these equal, because it is only comparing `firstName` and `lastName`, which are equal in `jason` and `anotherJason`. The `numChildren` property is completely ignored, resulting in `equals()` reporting true when it likely should report false.

DATA CLASSES ARE BEST LEFT ALONE

You can do a lot of things with data classes beyond the simple creation and declaration of properties. Listing 8.15 is a data class in its simplest form.

LISTING 8.15: A User data class in its simplest form

```
package org.wiley.kotlin.user

data class User(val email: String, val firstName: String, val lastName: String)
```

This couldn't be simpler, and in many ways, it is the epitome of a data class. There's a class with a few properties, and Kotlin is left to generate every bit of code associated with this class.

Compare this with Listing 8.16, a much more complicated version of a data class.

LISTING 8.16: A data class with less code to generate

```
package org.wiley.kotlin.person

data class Person(var firstName: String, var lastName: String) {
    var numChildren: Int = 0
    var fullName: String

    init {
        fullName = "$firstName $lastName"
    }

    override fun toString(): String {
        return fullName
    }
}
```

Here, there are additional properties (`numChildren` and `fullName`), an `init` block, and a `toString()` that prevents a generated version from being created. You could even have this extend a class to further complicate this class.

Ultimately, though, data classes actually start to *lose* value as you add customizations. They're at their best and most useful when you as a programmer let Kotlin do as much work as possible. Further, the more you start to override generated code, the more you pick up responsibilities in keeping the class consistent and the less it becomes a simple, "managed by Kotlin" helper.

In many ways, a data class is a bit like the default versions of `equals()` and `hashCode()` that you get in subclassing `Any`. Data classes (and the default implementations of `equals()` and `hashCode()`) are great if you can use them "as is." But remember: when you change `equals()`, you also have to change `hashCode()`, and quickly, you've got a lot more to manage. In the same way, as soon as you start changing a data class, you lose a lot of the advantages and flexibility that you've gained.

Your best bet is to simply declare your class, and if at all possible, leave it at that. Just because you *can* start overriding functions and adding behavior doesn't mean that you always should. Additionally, if you do feel that you *have* to add functionality or properties outside of the constructor or prevent functions from being generated, just take a minute and make sure that you are getting the most out of your data classes. You'll often find that there are ways to keep the class simple, and that simplicity makes the data class valuable.

Now that you've seen how Kotlin makes data handling easy with data classes, you're ready to look at some more built-in Kotlin helpers: constants, enums, and sealed classes.

Enums and Sealed, More Specialty Classes

STRINGS ARE TERRIBLE AS STATIC TYPE REPRESENTATIONS

It's time to look at one of the most common, and often annoying, aspects of building out class structures: constants and variables that use those constants. More specifically, the common problem of ensuring that a particular property can only have a certain set of values.

Listing 9.1 shows a very simple test class that will quickly illustrate this issue.

LISTING 9.1: Creating a user of a certain type

```
import org.wiley.kotlin.user.User

fun main() {
    val bardia = User("bardia@gmail.com","Bardia", "Dejban", "ADMIN")

    if (bardia.isAdmin()) {
```

continues

LISTING 9.1 *(continued)*

```
        println("You are frothing with power!")
    } else {
        println("Keep at it, you'll be an admin one day!")
    }
}
```

> **NOTE** *If you want to match the example files available for download, or just follow along exactly, name this class* EnumApp.

Pair this class with Listing 9.2, which slightly advances the User class from Chapter 8.

LISTING 9.2: A user class with a type

```
package org.wiley.kotlin.user

data class User(var email: String, var firstName: String, var lastName: String, var
type: String) {

    fun isAdmin(): Boolean {
        return type.equals("Admin")
    }

    override fun toString(): String {
        return "$firstName $lastName with an email of $email has a user
type of $type"
    }
}
```

Build and run this code, and you'll get a hint at the issue that enum classes—the subject of much of this chapter—address:

```
Keep at it, you'll be an admin one day!
```

Why does this user—given a type of "ADMIN"—not report back as an admin? Well, on its face, this is a capitalization issue. However, there's much more going on worth considering.

Strings Are Terrible Type Representations

At its heart, there are just a few important lines of code here. First, there's the instantiation of a new User instance:

```
val bardia = User("bardia@gmail.com","Bardia", "Dejban", "ADMIN")
```

Second, there's the code in the User class that actually reports whether a user is an admin or not:

```
fun isAdmin(): Boolean {
    return type.equals("Admin")
}
```

Of course, you can now quickly see why the bardia instance isn't reporting as an admin: bardia's type is "ADMIN" and the User class checks for the type being "Admin."

Capitalization Creates Comparison Problems

One easy fix is to take capitalization out of the picture. Make this change to isAdmin():

```
fun isAdmin(): Boolean {
    return type.equals("Admin", true)
}
```

This variant of equals() provides an additional parameter available to string comparisons: ignoreCase. By default, ignoreCase is false, and so "Admin" and "ADMIN" would not be considered equal. With the ignoreCase parameter set to true, "Admin" and "ADMIN" (as well as "aDMin" and "admin") all report as equal.

Now you can rerun the test class and you'll see that the bardia instance correctly reports as an admin:

```
You are frothing with power!
```

Before you get too comfortable though, this is still fragile. Seasoned developers know what comes next. Listing 9.3 adds another intended admin to the test class and runs through all the instances to check if they're admins.

LISTING 9.3: An updated test class that checks for admins

```
import org.wiley.kotlin.user.User

fun main() {
    val users = mutableListOf<User>()

    users.add(User("bardia@gmail.com","Bardia", "Dejban", "ADMIN"))
    users.add(User("shawn@gmail.com", "Shawn", "Khorrami", "Administrator"))

    for (user in users) {
        if (user.isAdmin()) {
            println("${user.firstName} ${user.lastName} is frothing with power!")
        } else {
            println("${user.firstName} ${user.lastName} is still working at it...
maybe one day!")
        }
    }
}
```

The output isn't too surprising:

```
Bardia Dejban is frothing with power!
Shawn Khorrami is still working at it... maybe one day!
```

The issue this time isn't capitalization, but using "Administrator" instead of some form of "Admin." The result is that "Admin" (the value used as the `true` value in User's `isAdmin()` method) doesn't match "Administrator," so the second instance reports back `false` from `isAdmin()`.

Again, this could be fixed with a tweak to the `isAdmin()` function:

```
fun isAdmin(): Boolean {
    return type.equals("Admin", true) ||
            type.equals("Administrator", true)
}
```

This sort of solution starts to get pretty silly fast, though. How far are you willing to go to catch every possible variation on "Admin"?

This Problem Occurs All the Time

This problem is a pretty common one in programming. As you build out different objects, whether they're users or people (the `Person` class used in earlier chapters), representations of cars or products, or even nontangible objects like transactions, you're going to need to classify instances of those objects. The property might not always be called `type`, as it was in `User`, but you'll often need a way to see if an instance is "this type of thing" or "that type of thing."

Put another way, this problem is bigger than dealing with string comparison. It's something that is pretty fundamental to programming and requires an elegant solution as you're going to have to implement that solution again and again.

String Constants Can Help . . . Some

A common "next step" in dealing with these types is to use constants. This is providing a fixed name that has a value, but then allowing other code to use that fixed name, and not worry about the value.

You can do this with a combination of the `const` keyword in Kotlin as a prefix to `val`. Listing 9.4 adds several constants to the `User` class.

LISTING 9.4: Adding a constant to the User class

```
package org.wiley.kotlin.user

data class User(var email: String, var firstName: String, var lastName: String, var
type: String) {
    const val ADMIN = "Admin"

    fun isAdmin(): Boolean {
        return type.equals(ADMIN)
    }

    override fun toString(): String {
```

```
        return "$firstName $lastName with an email of $email has a user
type of $type"
    }
}
```

To be clear, the `const` keyword doesn't change the type of the variable. In other words, take this line:

```
const val ADMIN = "Admin"
```

This line is *not* creating a different type than this:

```
val ADMIN = "Admin"
```

> **NOTE** *If you have experience programming in Java, you'll likely recognize this as the equivalent of a* `public static final` *variable. As you're about to see, though, the idea of* `static` *in Kotlin isn't represented in the same way that it is in Java.*

In either case, you end up with a `String`. That string could be "Admin" (as it is here) or "giraffe" or "asd8asd832nsa." But the comparison becomes easy now in `isAdmin()`:

```
fun isAdmin(): Boolean {
    return type.equals(ADMIN)
}
```

As long as the constant `ADMIN` is used on "both sides" of the comparison, this will work. You'll want to change the code in your test class to use this same constant:

```
users.add(User("bardia@gmail.com","Bardia", "Dejban", User.ADMIN))
```

Unfortunately, this is where you're going to hit a very strange wall. You'll know you've hit it when you get a compiler error like this:

```
Kotlin: Const 'val' are only allowed on top level or in objects
```

To fix this, you're going to have to take a step back from string comparison and learn about companion objects.

COMPANION OBJECTS ARE SINGLE INSTANCE

What's the problem with the code in Listing 9.5?

LISTING 9.5: Current User class . . . which won't compile

```
package org.wiley.kotlin.user

data class User(var email: String, var firstName: String, var lastName: String, var
type: String) {
    const val ADMIN = "Admin"
```

continues

LISTING 9.5 *(continued)*

```
    fun isAdmin(): Boolean {
        return type.equals(ADMIN)
    }

    override fun toString(): String {
        return "$firstName $lastName with an email of $email has a user
type of $type"
    }
}
```

The problem here is related to multiple instances of User existing.

Constants Must Be Singular

It's not obvious, but the problem here has to do with User being an object that can be (and must be) instantiated. As a reminder, here's a fragment of the test code using the User object:

```
users.add(User("bardia@gmail.com","Bardia", "Dejban", "ADMIN"))
users.add(User("shawn@gmail.com", "Shawn", "Khorrami", "Administrator"))
```

Once these two lines of code have executed, there are *two* instances of User. But *each* of those instances has a constant defined. The first instance, with a first name of "Bardia," has an ADMIN const, as does the second one, with a first name of "Shawn."

Kotlin recognizes this as a problem. The ADMIN constant doesn't exist once for all instances; it exists as a property of *each* User instance. For a String constant, that's not a huge deal. Suppose the first User instance was named bardia, and the second shawn. In that situation, you could compare bardia.ADMIN to shawn.ADMIN and get a true result. That's because String comparisons using equals() just compare the actual String values.

But if you were using == or any version of equals() that actually did a memory-level comparison, it's possible for a comparison of the same constant in two object instances to return false. That's a big problem.

> **NOTE** *For the Java programmers, this is exactly what the* static *keyword protects against. It tells Java to share a variable or method across all instances of the class. That's what you want in Kotlin but you need to learn how to do that the Kotlin way.*

On some level, the idea here is that bardia.ADMIN and shawn.ADMIN would not just return equal values, but that they would return the *same* value. And in this context, "the same" doesn't just mean that shawn.ADMIN.equals(bardia.ADMIN) would return true. It means more: that shawn.ADMIN and bardia.ADMIN are actually the same bit of memory, the same actual object at the very lowest level.

So take this one step further. For clarity, the goal is to actually reference User.ADMIN. That makes it much more obvious that there is a single ADMIN property shared across all instances of User.

Companion Objects Are Singletons

There is an important concept you need to understand now: that of a singleton. A *singleton* is a class that only has a single instance in a particular system. So even though you might reference that object—the singleton—in two or five of one hundred places in your code, there is only a single instance used in every reference.

In Kotlin, this is the role of a companion object. A companion object is declared inside the object it is a companion to. Listing 9.6 takes the code from Listing 9.5 and introduces a companion object, and then places the ADMIN definition within that companion object.

LISTING 9.6: Adding an unnamed companion object to User

```
package org.wiley.kotlin.user

data class User(var email: String, var firstName: String, var lastName: String, var
type: String) {
    companion object {
        const val ADMIN = "Admin"
    }

    fun isAdmin(): Boolean {
        return type.equals(ADMIN)
    }

    override fun toString(): String {
        return "$firstName $lastName with an email of $email has a user
type of $type"
    }
}
```

In this case, the companion object has no name, so you can reference the ADMIN property through User:

```
println(User.ADMIN)
```

So now you can go back to your test class and change your code to use this companion object property:

```
users.add(User("bardia@gmail.com","Bardia", "Dejban", User.ADMIN))
users.add(User("shawn@gmail.com", "Shawn", "Khorrami", User.ADMIN))
```

This code now compiles, and gives you the expected output:

```
Bardia Dejban is frothing with power!
Shawn Khorrami is frothing with power!
```

It's worth walking through exactly what's happening here to be really clear about the effect of a singleton.

First, two instances of User are created.

```
users.add(User("bardia@gmail.com","Bardia", "Dejban", User.ADMIN))
users.add(User("shawn@gmail.com", "Shawn", "Khorrami", User.ADMIN))
```

Both have as type `User.ADMIN`, which is the value defined in the companion object. So both have the exact same value for `type`. Again, that doesn't just mean that both have a `String` value of "ADMIN." It means that they both point to the same actual location in memory, which is "owned" by the `User`'s companion object.

Then, for each of these instances, the `isAdmin()` function is called, which runs this code:

```
fun isAdmin(): Boolean {
    return type.equals(ADMIN)
}
```

The `type` value is the `User.ADMIN` property, which is compared to the exact same property. And remember, this is actually going to result in memory equality, because every time you see `User.ADMIN`, or just `ADMIN` within the `User` class, the *exact same single instance object property* is being referenced.

That's the power and use of a companion object that's a singleton: it removes any ambiguity at all about a value, and ensures equality of comparisons at the deepest, lowest level possible.

Companion Objects Are Still Objects

Before returning to the problem at hand—constants and type safety for the `User` class and in general—it's worth looking a bit more at companion objects. At their heart, companion objects are singletons, but they're also just objects. They can have names, their own properties, and functions.

Listing 9.7 takes the version of `User` in Listing 9.6 and adds quite a bit to its companion, including a name. It then adds some functions to that object.

LISTING 9.7: Filling out the companion object for User

```
package org.wiley.kotlin.user

data class User(var email: String, var firstName: String, var lastName: String, var
type: String) {
    companion object Factory {
        const val ADMIN = "Admin"
        const val MANAGER = "Manager"
        const val END_USER = "End User"
        const val DEFAULT = END_USER

        fun createAdmin(email: String, firstName: String, lastName: String): User {
            return User(email, firstName, lastName, ADMIN)
        }

        fun createManager(email: String, firstName: String, lastName:
String): User {
            return User(email, firstName, lastName, MANAGER)
        }
    }

    fun isAdmin(): Boolean {
        return type.equals(ADMIN)
    }
}
```

```
        override fun toString(): String {
            return "$firstName $lastName with an email of $email has a user
    type of $type"
        }
    }
```

A couple of things to note here, all important as you begin using companion objects:

➤ A companion object can have a name, in this case, `Factory`.

➤ A companion object can have functions.

➤ Companion object functions often return the type of the containing object (in other words, the object this companion is the companion *to*).

This is a relatively clear addition. The companion object is now called `Factory`, and in addition to the `ADMIN` constant, it now defines a few additional user types: `MANAGER` and `END_USER`. It also defines a `DEFAULT` type, which references the `END_USER` type.

It also adds some functions that create a couple of user types with just an email, first name, and last name. This means that the users of this code (not `User` instances, but actual developers) don't need to keep up with constants like `User.ADMIN` if they don't want to. Instead, they can call `createAdmin()` or `createManager()` and the details of types will get taken care of by the `Factory` companion object.

You Can Use Companion Objects without Their Names

If you updated your code to match Listing 9.7 and then compiled or ran your project, you might have noticed something pretty unusual. These lines of code did *not* result in a compiler error:

```
    users.add(User("bardia@gmail.com","Bardia", "Dejban", User.ADMIN))
    users.add(User("shawn@gmail.com", "Shawn", "Khorrami", User.ADMIN))
```

You would expect all the references to `User.ADMIN` to change to `User.Factory.ADMIN`. But that's another wrinkle (or convenience) that companion objects add: you *can* use the companion object's name, but you don't have to. You can reference properties and functions as if they belonged to the companion class.

Using a Companion Object's Name Is Optional

You would expect that you'd need to modify these lines to look like this:

```
    users.add(User("bardia@gmail.com","Bardia", "Dejban", User.Factory.ADMIN))
    users.add(User("shawn@gmail.com", "Shawn", "Khorrami", User.Factory.ADMIN))
```

But that's not the case, and it makes using companion objects quite simple. You can also now use the helper functions, as shown in Listing 9.8.

LISTING 9.8: Using helper functions in the test class

```
import org.wiley.kotlin.user.User

fun main() {
```

continues

LISTING 9.8 *(continued)*

```
    val users = mutableListOf<User>()

    users.add(User("bardia@gmail.com","Bardia", "Dejban", User.ADMIN))
    users.add(User("shawn@gmail.com", "Shawn", "Khorrami", User.ADMIN))

    users.add(User.createAdmin("bahar@gmail.com", "Bahar", "Dejban"))
    users.add(User.createManager("bastion@email.com", "Bastion", "Fennell"))

    for (user in users) {
        if (user.isAdmin()) {
            println("${user.firstName} ${user.lastName} is frothing with power!")
        } else {
            println("${user.firstName} ${user.lastName} is still working at it...
maybe one day!")
        }
    }
}
```

This does make the code a little simpler and is also more self-documenting.

> **NOTE** *Self-documenting is a term that generally means code is written in such a way that it's obvious what it does from just reading it—even if you're not familiar with the code. In this case, the function name* createAdmin *clearly creates an admin, and the same is true for* createManager. *Both of those function names are self-documenting. You can also create variables that are named cryptically, like _ a and u2, or that are also self-documenting, like* index *and* user2. *It's always better to write self-documenting code when possible.*

Note that Listing 9.9 is a slightly different version of Listing 9.8 that uses the same functions, but this time introduces the companion object name.

LISTING 9.9: Using helper functions and adding the companion object name

```
import org.wiley.kotlin.user.User

fun main() {
    val users = mutableListOf<User>()

    users.add(User("bardia@gmail.com","Bardia", "Dejban", User.ADMIN))
    users.add(User("shawn@gmail.com", "Shawn", "Khorrami", User.ADMIN))

    users.add(User.Factory.createAdmin("bahar@gmail.com", "Bahar", "Dejban"))
    users.add(User.Factory.createManager("bastion@email.com", "Bastion",
"Fennell"))
```

```
    for (user in users) {
        if (user.isAdmin()) {
            println("${user.firstName} ${user.lastName} is frothing with power!")
        } else {
            println("${user.firstName} ${user.lastName} is still working at it...
maybe one day!")
        }
    }
}
```

These two bits of code—Listing 9.8 and Listing 9.9—are exactly the same when executed. So then the eternal question that plagues anxious developers is: if both are OK, which is *better*?

Using a Companion Object's Name Is Stylistic

Honestly, neither is better. But there's an argument to be made that adding `Factory` makes the code a little clearer, a little more self-documenting. It becomes really obvious that the `createAdmin()` and `createManager()` functions are factory functions, and also likely singletons, because of the nature of factory methods in general.

> **NOTE** *It's common, and even expected, that factory methods are generally written as singletons, as they don't typically maintain state. They exist to do one thing: create object instances.*

So adding the `Factory` companion object name is a little clearer. Of course, that only helps if the companion object is named `Factory`, or something else helpful (maybe `UserCreator` or `UserFactory`, although `Factory` on its own is pretty good).

But also note that it really *doesn't* help much to add `Factory` in when referencing the constants in that companion object. This code is pretty clear:

```
users.add(User("bardia@gmail.com","Bardia", "Dejban", User.ADMIN))
users.add(User("shawn@gmail.com", "Shawn", "Khorrami", User.ADMIN))
```

This code actually introduces the companion object name without much value:

```
users.add(User("bardia@gmail.com","Bardia", "Dejban", User.Factory.ADMIN))
users.add(User("shawn@gmail.com", "Shawn", "Khorrami", User.Factory.ADMIN))
```

So it can become a little bit of a mixed bag; sometimes you may use the companion object name, and sometimes you may not. It really becomes a matter of style and choice. Again, if you simply choose the option in a given situation that makes for the clearest code, you're usually making a good choice—even if you're not totally consistent.

Companion Object Names Are Hard

It's worth pointing out that getting just the right companion name is often really hard . . . or it's really easy. If that sounds odd, think about a couple of things that are in play when naming your companion object:

➤ Any object (like User) can only have *one* companion object.

➤ Anything that should be single-instance, shared across instances, or a singleton must be in that one companion object.

➤ Often, high-level classes (again, like User, that may have multiple subclasses) need to share functionality (like factory methods) and constants (like User.MANAGER and User.ADMIN).

If you don't have to represent both constants and factory methods, naming is easy. Listing 9.10 shows a version of User with just constants, uses a companion object named Constants, and results in a very clear, self-documenting object.

LISTING 9.10: The User object with just constants in the companion object

```
package org.wiley.kotlin.user

data class User(var email: String, var firstName: String, var lastName: String, var
type: String) {
    companion object Constants {
        const val ADMIN = "Admin"
        const val MANAGER = "Manager"
        const val END_USER = "End User"
        const val DEFAULT = END_USER
    }

    fun isAdmin(): Boolean {
        return type.equals(ADMIN)
    }

    override fun toString(): String {
        return "$firstName $lastName with an email of $email has a user
type of $type"
    }
}
```

That's easy. But as soon as you add in the factory methods (like Listing 9.7 earlier), then neither Constants nor Factory works for everything in the companion object.

So, what do you do? Well, you can avoid the problem altogether, as you'll see next, and later in the chapter, you'll see a way to handle constants in yet another manner and avoid the problem in a different way.

You Can Skip the Companion Object Name Altogether

If you look all the way back at Listing 9.6, you'll see that companion objects don't *have* to have names. In fact, Listing 9.11 has all the constants and helper functions that you've added to User's companion object, but this time without a companion object name.

LISTING 9.11: User with an unnamed companion object

```
package org.wiley.kotlin.user

data class User(var email: String, var firstName: String, var lastName: String, var
type: String) {
    companion object {
        const val ADMIN = "Admin"
        const val MANAGER = "Manager"
        const val END_USER = "End User"
        const val DEFAULT = END_USER

        fun createAdmin(email: String, firstName: String, lastName: String): User {
            return User(email, firstName, lastName, ADMIN)
        }

        fun createManager(email: String, firstName: String, lastName:
String): User {
            return User(email, firstName, lastName, MANAGER)
        }
    }

    fun isAdmin(): Boolean {
        return type.equals(ADMIN)
    }

    override fun toString(): String {
        return "$firstName $lastName with an email of $email has a user
type of $type"
    }
}
```

> **NOTE** *If you're following along, you may get some compilation errors after making the change shown in Listing 9.11. You'll need to remove any references to* Factory *in your test class to get rid of those issues.*

This is a sort of "best of both worlds" approach, although it also could be seen as a "worst of both worlds" approach, too. On the upside, you no longer have constants in a companion object called Factory, or factory methods in a companion object called Constants. You also don't have to decide in your classes that use the object whether to add the companion object name.

On the downside, you're getting less specificity, and it's easy to throw anything into the companion object. A name like `Constants` or `Factory` gives at least subtle guidance; a nameless companion object does not.

Regardless, this is an option, and does provide consistency and some naming benefits.

ENUMS DEFINE CONSTANTS AND PROVIDE TYPE SAFETY

There remains a different path for constants, though. First, you should recognize that even with a companion object—whether it's called `Factory` or `Constants`, or is nameless—there is still a potential for misuse. Look at the declaration for the `User` class's primary constructor:

```
data class User(var email: String, var firstName: String, var lastName: String, var
type: String) {
```

Now, even with constants like `User.ADMIN` and `User.MANAGER`, you're still just passing in a `String`. `User.ADMIN` is a `String` ("Admin") and `User.MANAGER` is a `String` ("Manager").

Here's how you'd *prefer* users work with the `User` class:

```
users.add(User("bardia@gmail.com","Bardia", "Dejban", User.ADMIN))
users.add(User("shawn@gmail.com", "Shawn", "Khorrami", User.ADMIN))
```

But here's another perfectly legal bit of code:

```
users.add(User("bardia@gmail.com","Bardia", "Dejban", "AdMIN"))
users.add(User("shawn@gmail.com", "Shawn", "Khorrami", "Administrator"))
```

So all the work to define constants and use those constants in your `equals()` method really doesn't provide any type safety. You still are allowing any `String` as input, and that `String` might match your constants, it might not, and it might even be mistyped or miscapitalized or complete gibberish.

You want more than just some predefined `String` values. You actually want full-blown type safety, where you define a new type—in this case, something like `UserType`—and limit the available types. Your `User` instances then *only* accept the types defined by `UserType`.

Enums Classes Provide Type-Safe Values

An enum is a special type of class that provides exactly what you need in this situation: specific values and a new type that can be used in your own classes.

Listing 9.12 is a simple enum that defines a new type: `UserType`. You just preface the `class` keyword with `enum`, in the same way that you prefaced the `class` keyword with `data` for data classes.

LISTING 9.12: A new UserType enum

```
package org.wiley.kotlin.user

enum class UserType {
    ADMIN,
```

```
    MANAGER,
    END_USER
}
```

This is an incredibly simple class that is going to provide a lot more in terms of type safety than `String` constants. You can access the constants by the class name and the constant name, just as you have been doing: `UserType.ADMIN`, `UserType.MANAGER`, `UserType.END_USER`.

The key difference in this and companion objects is that the `String` constants in a companion object are `String` types, and here, you've defined a new type, `UserType`.

That means you can update `User` to require `UserType` as the input type for the type property, as well as update the `equals()` method. Finally, you can remove the constants from the companion object. The result is shown in Listing 9.13.

LISTING 9.13: Using the UserType enum in User

```
package org.wiley.kotlin.user

data class User(var email: String, var firstName: String, var lastName: String, var
type: UserType) {
    companion object {
        const val ADMIN = "Admin"
        const val MANAGER = "Manager"
        const val END_USER = "End User"
        const val DEFAULT = END_USER

        fun createAdmin(email: String, firstName: String, lastName: String): User {
            return User(email, firstName, lastName, UserType.ADMIN)
        }

        fun createManager(email: String, firstName: String, lastName:
String): User {
            return User(email, firstName, lastName, UserType.MANAGER)
        }
    }

    fun isAdmin(): Boolean {
        return type.equals(UserType.ADMIN)
    }

    override fun toString(): String {
        return "$firstName $lastName with an email of $email has a user
type of $type"
    }
}
```

You've now removed any stray types from being used and removed the possibility of a random `String` being supplied by other code.

Listing 9.14 shows an updated version of the test class that takes advantage of the `UserType` enum.

LISTING 9.14: Creating users with a specific UserType

```
import org.wiley.kotlin.user.User
import org.wiley.kotlin.user.UserType

fun main() {
    val users = mutableListOf
    users.add(User("bardia@gmail.com","Bardia", "Dejban", UserType.ADMIN))
    users.add(User("shawn@gmail.com", "Shawn", "Khorrami", UserType.ADMIN))

    users.add(User.createAdmin("bahar@gmail.com", "Bahar", "Dejban"))
    users.add(User.createManager("bastion@email.com", "Bastion", "Fennell"))

    for (user in users) {
        if (user.isAdmin()) {
            println("${user.firstName} ${user.lastName} is frothing with power!")
        } else {
            println("${user.firstName} ${user.lastName} is still working at it...
maybe one day!")
        }
    }
}
```

Creating a User with a type is shown, and the factory methods on the User's companion object still work with the changes introduced in Listing 9.13.

Enums Classes Are Still Classes

Although enums typically are used to provide you type-safe constants, they are still classes. That means that they can override methods and define behavior. For example, suppose you want to define a custom implementation of toString(). You can do that, as shown in Listing 9.15.

LISTING 9.15: Overriding a method in an enum

```
package org.wiley.kotlin.user

enum class UserType {
    ADMIN,
    MANAGER,
    END_USER;

    override fun toString(): String {
        return "User type of ${name}";
    }
}
```

It's important to note that this looks like any other class that overrides behavior from the implicitly inherited Any class. But there is one wrinkle: the use of a property called name.

Enums Give You the Name and Position of Constants

name is a property you get in all enums. It will report the specific name of the object chosen, so in this case, ADMIN or MANAGER or END_USER. So you might use that like this:

```
println(user.type);
```

You'd get something back like:

```
User type of MANAGER;
```

You also get access to ordinal, which will give you the numeric index of the constant you've chosen. This seems odd until you think about an enum as essentially a list of constant objects.

So for UserType, this is the simple declaration without any overridden functions:

```
enum class UserType {
    ADMIN,
    MANAGER,
    END_USER;
}
```

ordinal for ADMIN is 0, MANAGER would report 1, and END_USER 2.

You won't use ordinal as often, but name is quite helpful.

Each Constant in an enum Is an Object

In addition to an enum itself being a class, each constant is basically an anonymous object. This means that they, too, can define functions and behavior.

Listing 9.16 adds a custom toString() implementation for just the ADMIN constant.

LISTING 9.16: Overriding a function in a particular constant object

```
package org.wiley.kotlin.user

enum class UserType {
    ADMIN {
        override fun toString(): String {
            return "Administrator"
        }
    },
    MANAGER,
    END_USER;

    override fun toString(): String {
        return "User type of ${name}"
    }
}
```

Now try this code snippet:

```
println(UserType.ADMIN)
println(UserType.MANAGER)
```

The first line prints:

```
Administrator
```

This is using the overridden version of toString() specific to UserType.ADMIN.

The second line prints:

```
User type of MANAGER
```

This is using the version of toString() overridden by the enum, but nothing specific to MANAGER (other than the name property of that constant).

Each Constant Can Override Class-Level Behavior

Although you've not dealt much with abstract definitions yet (they're coming soon!), consider Listing 9.17.

LISTING 9.17: Defining a function to be overridden by constant objects

```kotlin
package org.wiley.kotlin.user

enum class UserType {
    ADMIN {
        override fun toString(): String {
            return "Administrator"
        }

        override fun isSuperUser() = true
    },
    MANAGER {
        override fun isSuperUser() = false
    },
    END_USER {
        override fun isSuperUser() = false
    };

    override fun toString(): String {
        return "User type of ${name}";
    }

    abstract fun isSuperUser(): Boolean
}
```

Here, the enum defines a function—isSuperUser()—and declares it as abstract. This requires the constant objects defined in the enum to implement that function, which each does.

So now you can call isSuperUser() on any UserType instance:

```kotlin
if (users.first().type.isSuperUser()) {
    println("The first user has super user access")
}
```

At this point, you have a pretty extensive and configurable tool—enums—for representing constant values. You don't have to rely on constants that don't provide type safety, and you don't have to worry about naming of companion objects.

You also get actual classes, with constant objects (not strings or other data types) that can define their own behavior and even override functions up the inheritance tree.

But what if you want more than a constant? What if you want this same level of type safety *and* you want it in a class hierarchy? This is where sealed classes come in.

SEALED CLASSES ARE TYPE-SAFE CLASS HIERARCHIES

Suppose you wanted to model luxury automobiles in Kotlin. Maybe you're building a site that allows luxury cars to be traded in through an online process, and you need an object-based system for representing the different types of cars that could be traded.

You might start with an `Auto` class, something very simple, like Listing 9.18.

LISTING 9.18: A very simple Auto class

```
package org.wiley.kotlin.auto

data class Auto(var make: String, var model: String) {

}
```

This class is a data class, which is great. You're probably also thinking that it could use an enum, and that's true. You might add to it an enum for popular makes, like that shown in Listing 9.19.

LISTING 9.19: The beginning of an enum with makes of automobiles

```
package org.wiley.kotlin.auto

enum class Make {
    ALFA_ROMEO,
    AUDI,
    BMW,
    CADILLAC,
    JAGUAR,
    LAND_ROVER,
    LEXUS,
    MASERATI,
    MERCEDES
}
```

> **WARNING** *For the luxury car fans out there, this* Auto *enum is woefully incomplete, and a lot of time could be spent debating which makes belong and which don't. Please take this list as representative, and not exhaustive!*

Now you could change the primary constructor to Auto to use this enum for the make:

```
data class Auto(var make: Make, var model: String) {
```

This doesn't change much about Auto, but it does now ensure type safety, and also that an Auto instance doesn't have a make that doesn't exist, is misspelled, or isn't included in the luxury list. So far, it seems that using enums and data classes is sufficient.

Enums and Class Hierarchies Work for Shared Behavior

At this point, you could build out Auto with a drive() function, perhaps a featureList() function that returns that model's features, and a whole array of additional behavior. Then, you can build out the behavior for the class. Some of it will be shared—you really don't need a drive() function that's different for each model type—and some might exist on the constant objects in Model.

You could also model this with a class hierarchy. Instead of using the Make enum, you might create a base class called Auto, and then extend that class with a Lexus and Jaguar and Maserati subclass. Each would extend Auto, and each would override any needed behavior.

There's nothing wrong with this approach. In fact, it's likely the *correct* approach, because it accurately models the hierarchy between a more basic type, Auto, and more specific types, Lexus, Jaguar, and others. The base type has some shared behavior, like drive() and toString(), and the subclasses add or override Auto with behavior that is specific to that subclass.

Additionally, you can add additional subclasses. When the Genesis make came out, you'd just add that as a new subclass, or as a new constant in the Make enum.

But there are some cases where this sort of approach doesn't work well, and that's where sealed classes come in.

Sealed Classes Address Fixed Options and Non-Shared Behavior

It's time to revisit your high school mathematics course. Remember those days? You had a set of values and then a set of operators, and those operators could be applied to values. The basic operators were addition, subtraction, multiplication, and division. That's really it. There were some more—exponents and roots—that are less common but bring the total to six.

Once you knew those six operations, you were more or less done. That was it. The usage got more complex, but the operators stayed basically the same. So suppose you wanted to build out a calculator program in Kotlin. You'd need to represent these operations.

> **NOTE** *If you spend much time reading up on Kotlin online or in books, you're going to find that almost every time sealed classes come up, the examples used are centered on operations. It's a perfect example for sealed classes, and it turns out that there aren't a lot of different good sealed class examples.*
>
> *You can build sealed classes in other ways; it just turns out that in most of those cases, a sealed class isn't really needed. So rather than make up an example that really doesn't lend itself to a sealed class, we're taking the same approach: using a perfect example of when a sealed class is a good idea, even if it is the same example you'd find in other resources.*

Naturally, you'd start with a base class, perhaps called `Operation`. This is the analog to `Auto` in the prior examples; everything would then inherit from `Operation`. `Operation` isn't an enum, either. You don't want constants for `Divide` or `Add` or the others; you want actual subclasses with behavior.

So you'd likely have something like Listing 9.20 for `Operation`.

LISTING 9.20: A simple Operation class built to be subclassed

```
package org.wiley.kotlin.math

open class Operation {

}
```

Listing 9.21 shows an empty inheriting class, `Add`.

LISTING 9.21: An empty Add class that subclasses Operation

```
package org.wiley.kotlin.math

class Add : Operation() {

}
```

Sealed Classes Don't Have Shared Behavior

There's not much interesting here at first glance. But the key is considering the shared behavior of the different operations. What is shared between `Add`, `Subtract`, `Multiply`, `Divide`, and the other operations?

Asked another way, what functions would be defined in `Operation` that would apply across all operations? Are there any? Even one?

The answer is, "Not really." Each operation is indeed an operation, but they all operate in fundamentally different ways. This is one of two indications that you might have a good use case for a sealed class.

> **Criterion 1:** A sealed class represents common subclasses that do not share behavior.

Sealed Classes Have a Fixed Number of Subclasses

The second criterion is a little more nuanced. Put simply, it's:

> **Criterion 2:** A sealed class represents a class hierarchy where the options are fixed, and an object can only be one of those options.

This is nuanced because at first glance, any class hierarchy meets this criterion. Consider, for example, the `Auto` hierarchy. There are only a certain number of subclasses (`Maserati`, `Mercedes`, and so forth), and an `Auto` instance can only be one of those. But the nuance is in this: that list of subclasses is fixed only because those are the only subclasses defined, and not because the list of subclasses is actually exhaustive. As pointed out earlier, there may later be additional subclasses added.

This is quite different from `Operation`: the possible subclasses really are fixed, and known. They aren't going to grow over time (unless someone comes up with a new way to combine two numbers algebraically!) or shift.

In this case, you'd declare the sealed class with the `sealed` keyword where you'd normally see `open` or `data`. Listing 9.22 defines a new sealed class, `Operation`.

LISTING 9.22: Operation as a sealed class

```
package org.wiley.kotlin.math

sealed class Operation {

}
```

> **WARNING** *If you're following along, you'll need to replace the code shown in Listing 9.20 with Listing 9.22. Kotlin (and almost any language) won't allow you to define two different* Operation *classes in the same package.*

If you created the `Add` class shown earlier in Listing 9.21, you'll need to remove that file altogether, as Kotlin will otherwise report this error:

```
Kotlin: Cannot access '<init>': it is private in 'Operation'
```

This is because sealed classes aren't subclassed in the way you're used to subclassing. Instead, they appear within the sealed class file. Listing 9.23 shows a compilable version of `Operation`.

Listing 9.23 Operation as a sealed class with subclasses

```
package org.wiley.kotlin.math

sealed class Operation {
    class Add(val value: Int) : Operation()
    class Subtract(val value: Int) : Operation()
    class Multiply(val value: Int): Operation()
    class Divide(val value: Int): Operation()
    class Raise(val value: Int): Operation()
}
```

You can now compile this (after removing `Add`) without error.

> **NOTE** *If you're wondering what happened to taking the root of something, it's represented by* `Raise`. `Raise` *in this model actually handles exponentiation (5 raised to the 2nd power is 25) as well as roots (9 raised to ½ is actually the same as the square root of 9, which is 3). This approach keeps the code a little simpler, and since we're not actually building a complete calculator here, it also keeps the focus on the concepts.*

Subclasses of a Sealed Class Don't Always Define Behavior

Listing 9.23 should strike you as a bit odd. While there are subclasses, those subclasses don't define any behavior. They do each define a primary constructor—each subclass takes in an `Int` value—but no behavior.

That's OK; you typically don't use these classes in the way you'd use a class modeling an object. This is also where the all-important `when` control structure reappears, with new wrinkles.

when Requires All Sealed Subclasses to Be Handled

You've seen `when` more than a few times, often used like this:

```
when {
    (anotherNumber % 2 == 0) -> println("The number is divisible by 2!")
    (anotherNumber % 3 == 0) -> println("The number is divisible by 3!")
    (anotherNumber % 5 == 0) -> println("The number is divisible by 5!")
    (anotherNumber % 7 == 0) -> println("The number is divisible by 7!")
    (anotherNumber % 11 == 0) -> println("The number is divisible by 11!")
    else -> println("This thing is hardly divisible by anything!")
}
```

This is perfectly legitimate Kotlin, and a great use of `when`. However, the rules change when you're working with a sealed class. To understand this, you'll need to work backward a bit.

First, suppose the goal is to perform an operation like this:

```
5 + 4
```

You obviously are looking to get back a result of 9 here. This requires using the `Add Operation`, with a value of 4:

```
Add(4)
```

So we really want to create a mechanism of passing in a value (like 5) and an `Operation` (like `Add`) that also has a value (`Add(4)`).

The obvious choice here is a function. Let's call that function `execute()`, and it would take in those things: an input value and an `Operator` instance that has its own value.

Go ahead and create a new class called `Calculator` and enter in the code shown in Listing 9.24.

LISTING 9.24: A very simple calculator with one function

```
package org.wiley.kotlin.math

class Calculator {
    fun execute(input: Int, operation: Operation) {
        when (operation) {
            is Operation.Add          -> input + operation.value
        }
    }
}
```

So now you would (theoretically) be able to call `execute(5, Operation.Add(4))` and get back 20.

However, if you try to compile `Calculator`, you're going to get a rather odd error. It will look something like this:

```
Kotlin: 'when' expression must be exhaustive, add necessary 'is Subtract', 'is
Multiply', 'is Divide', 'is Raise' branches or 'else' branch instead
```

when Expressions Must Be Exhaustive for Sealed Classes

The key is that the `when` expression must be "exhaustive." This is what a sealed class introduces, and, if you recall, was one of the two criteria for a sealed class.

The `when` expression uses criterion 2, knowing that the options in a sealed class are fixed, and requires that every one of those subclasses is represented. Right now, only the `Add` operation is covered, leaving `Subtract`, `Multiply`, `Divide`, and `Raise` missing.

This is easy enough to fix, and a solution is shown in Listing 9.25.

LISTING 9.25: Adding additional sealed class subclasses to the when expression

```
package org.wiley.kotlin.math

import kotlin.math.pow
```

```
class Calculator {
    fun execute(input: Int, operation: Operation): Int =
        when (operation) {
            is Operation.Add         -> input + operation.value
            is Operation.Subtract    -> input - operation.value
            is Operation.Multiply    -> input * operation.value
            is Operation.Divide      -> input / operation.value
            is Operation.Raise       ->
(input.toFloat().pow(operation.value.toFloat())).Int()
        }
}
```

> **NOTE** *There's not a lot of time spent on the details of how this* when *and* execute() *function work because it's really not critical to understanding the current topic: sealed classes. That said, for the first few options, the function just takes the correct operation in Kotlin. For* Raise, *the* kotlin.math.pow() *function is used, but since it both takes and returns only* Float *or* Double, *there's some extra conversion involved.*

This class will now compile because every subclass of the sealed class is covered.

There's also a pretty key exception to the "every subclass must be covered" rule, and it's a subtle one. Note that in Listing 9.25, the when evaluates and the result is directly returned. In this case, it's returned as the return value of execute(). That's why after the function definition for execute(), it has an = sign before the when.

That's the rule: *if* the when evaluates and is the right-hand side of an expression or returned from a function, the when must handle every subclass. If the when is *not* the right-hand side of an expression or returned, the rule doesn't apply. So this code is perfectly legal:

```
class Calculator {
    fun execute(input: Int, operation: Operation): Int {
        when (operation) {
            is Operation.Add -> input + operation.value
            is Operation.Subtract -> input - operation.value
        }

        return 0
    }
}
```

Obviously, the code doesn't make much sense in that it always returns 0, but it does make the point: the when is not subject to the "every subclass" rule since it's no longer returned from execute().

```
class Calculator {
    fun execute(input: Int, operation: Operation): Int {
        val returnVal : Int = when (operation) {
            is Operation Add -> input + operation.value
```

```
            is Operation.Subtract -> input - operation.value
            is Operation.Raise -> (input.toFloat().pow(operation.value.
toFloat())).toInt()
            }

        return returnVal;
    }
}
```

In this case, the rule applies because the `when` is the right-hand side of an assignment to `returnVal`. The `when` is missing the `Operation.Multiply` and `Operation.Divide` subclasses, so an error results at compilation.

It's important to note that if you *did* add another subclass to `Operation`, every `when` you have is going to immediately break (unless that `when` uses an `else`, which is covered next).

else Clauses Usually Don't Work for Sealed Classes

It is technically legal to use an `else` clause in your `when` to ensure that all possible subclasses of a sealed class are handled. Listing 9.26 shows a version of `Calculator` that uses an `else` and is perfectly legal Kotlin code.

LISTING 9.26: Adding an else to a when

```
package org.wiley.kotlin.math

import kotlin.math.pow

class Calculator {
    fun execute(input: Int, operation: Operation): Int {
        val returnVal : Int = when (operation) {
            is Operation.Add -> input + operation.value
            is Operation.Subtract -> input - operation.value
            // is Operation.Multiply -> input * operation.value
            // is Operation.Divide -> input / operation.value
            is Operation.Raise -> (input.toFloat().pow(operation.value.
toFloat())).toInt()
            else -> throw Exception("Operation not yet supported")
            }

        return returnVal;
    }
}
```

Generally speaking, the code in Listing 9.26 turns out to be almost the *only* sort of `else` that makes sense for a sealed class. The `else` is a true catch-all, but rather than invoking shared behavior for multiple subclasses of `Operation`, it throws an `Exception` to indicate that the functionality for the `Operation` subclasses not covered explicitly hasn't been written yet, or isn't supported in some way.

In other words, this is a rather unusual case where multiple subclasses of a sealed class (in this case, `Operation`) do share behavior, even if that behavior is that there's unimplemented behavior. But it calls out the point, especially if you go back to the criteria for sealed classes shown earlier.

Criterion 1 hammers the point home: sealed classes *do not share behavior*. They each operate independently, and the only thing they do share is a class hierarchy. Given that, it only makes sense that while you *can* use an else in a when that involves a sealed class, it rarely makes sense. An else by definition is "shared behavior for anything not explicitly handled," which goes against criterion 1 for sealed classes.

It is a better idea to write a when in the situation where you have clear, explicit behavior for each subclass of the sealed class.

else Clauses Hide Unimplemented Subclass Behavior

Take things a step further. It's actually a *bad* idea to use an else in a when with sealed classes. Take a look at Listing 9.27, which can turn into sneaky problems.

LISTING 9.27: Using an else to catch any extra subclasses

```kotlin
package org.wiley.kotlin.math

import kotlin.math.pow

class Calculator {
    fun execute(input: Int, operation: Operation): Int {
        val returnVal : Int = when (operation) {
            is Operation.Add -> input + operation.value
            is Operation.Subtract -> input - operation.value
            is Operation.Multiply -> input * operation.value
            is Operation.Divide -> input / operation.value
            is Operation.Raise -> (input.toFloat().pow(operation.value.
toFloat())).toInt()
            else -> {
                // Catch all
                0
            }
        }

        return returnVal;
    }
}
```

This is legal code, but generally considered a bad idea. This code hides any subclasses of Operation that might be added down the line. For example, suppose Operation was updated to look like Listing 9.28.

LISTING 9.28: Adding subclasses to Operation

```kotlin
package org.wiley.kotlin.math

sealed class Operation {
    class Add(val value: Int) : Operation()
    class Subtract(val value: Int) : Operation()
```

continues

LISTING 9.28 *(continued)*

```
        class Multiply(val value: Int): Operation()
        class Divide(val value: Int): Operation()
        class Raise(val value: Int): Operation()

        class And(val value: Int): Operation()
        class Or(val value: Int): Operation()
}
```

In this new situation, there are additional subclasses of `Operation`, `Add`, and `Or`. That would certainly imply that `Calculator` should now stop working and require some additional handling of these two operations. But if `Calculator` is coded as shown in Listing 9.27, it will happily continue to compile and be usable because of the `else`:

```
else -> {
    // Catch all
      0
}
```

Anytime the `Operation` is `And()` or `Or()`, a `0` will be returned. So not only is `Calculator` compiling when it likely shouldn't, it's now returning erroneous values.

There's really only one good generalized reason to implement an `else`: to indicate a missing subclass.

> **WARNING** *It's always a little dangerous to say things like "only one good reason" or "you never should do something." No sooner will the words come out than someone will have an unusual but perfectly legitimate counterexample. Hopefully the word "generalized" in "only one good generalized reason" will make the message clearer: in the general case, an* `else` *for a sealed class outside of generating an exception isn't good code.*

If you do want to have a class that compiles a `when` statement and is resistant to a sealed class adding new subclasses, you could go with something like Listing 9.29. Every subclass of `Operation` is handled, but the `else` covers any new subclasses that might be added down the line. However, it covers them by raising an exception; the code will compile but if a subclass not explicitly covered is passed in at runtime, the program will error out.

LISTING 9.29: Moving errors from compile-time to runtime

```
package org.wiley.kotlin.math

import kotlin.math.pow

class Calculator {
    fun execute(input: Int, operation: Operation): Int {
        val returnVal : Int = when (operation) {
            is Operation.Add -> input + operation.value
            is Operation.Subtract -> input - operation.value
```

```
            is Operation.Multiply -> input * operation.value
            is Operation.Divide -> input / operation.value
            is Operation.Raise -> (input.toFloat().pow(operation.value.
toFloat())).toInt()
            else -> {
                throw Exception("Unhandled Operation: ${operation}")
            }
        }
    }

    return returnVal;
    }
}
```

Compile the version of `Operation` shown in Listing 9.28 and then run the test program like this:

```
var calc = Calculator()
println(calc.execute(5, Operation.Add(4)))
```

This code compiles despite the `else` not explicitly handling the `And` or `Or` `Operation` subclasses. And this code will correctly return 9.

Now try this bit of code:

```
println(calc.execute(2, Operation.And(4)))
```

The result will be an error that looks a bit like this:

```
Exception in thread "main" java.lang.Exception: Unhandled Operation: org.wiley.
kotlin.math.Operation$And@506e6d5e
        at org.wiley.kotlin.math.Calculator.execute(Calculator.kt:14)
        at EnumAppKt.main(EnumApp.kt:32)
        at EnumAppKt.main(EnumApp.kt)

Process finished with exit code 1
```

This is a runtime error, though, not a compile-time one. It basically serves as a reminder to you that you need to handle a missing `Operation` subclass.

> **NOTE** *This section is focused on using an* `else` *with a* when *specifically for sealed classes. However, the point is valid for all* when *statements to a lesser degree. An* `else` *makes it easy to miss that a new subclass or matching expression could be introduced and not get caught explicitly by the* when.
>
> *While there are good reasons to use an* `else` *if you're not dealing with sealed classes, one of those good reasons is not as a generic catch-all. In fact, in cases where an* `else` *would work, consider listing all the possible matching expressions using an* or (||) *and avoiding the* `else` *altogether.*

At this point, you've seen a whole lot of special classes: data classes, enums, sealed classes, and the beginning of abstract classes and open classes, too. While there's still more to say on classes—there almost always is in an object-oriented language—it's time to focus in a little more and look at how functions are also quite detailed and flexible in Kotlin.

10

Functions and Functions and Functions

WHAT'S IN THIS CHAPTER?

- ➤ Functions that take in functions as arguments
- ➤ Functions that return functions
- ➤ Anonymous functions and lambda expressions
- ➤ Warnings about lambdas

REVISITING THE SYNTAX OF A FUNCTION

By now, even if this book is your very first introduction to Kotlin, you've seen functions hundreds of times. They are the building blocks of Kotlin, whether you're writing a simple data class or a sealed class with subclasses or a complicated custom class.

In this chapter, you're going to get deeper into functions: both the syntax that you've been using all along, and the number of ways you can expand and stretch that syntax to do much more interesting things.

Functions Follow a Basic Formula

Listing 10.1 is about the simplest Kotlin file possible. It's called *FunctionApp.kt*, and has only one thing in it: a simple function called `main()`.

LISTING 10.1: A super simple Kotlin class with a super simple function

```
fun main() {
    println("Here we go...")
}
```

Listing 10.1 is simple because it makes it easy to see the components of a function. Figure 10.1 breaks these down a bit further.

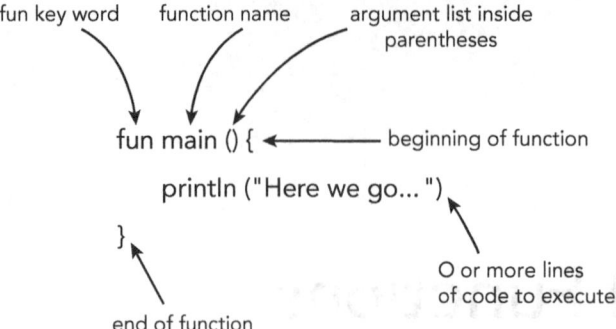

FIGURE 10.1: A function always has the same basic parts.

Every function has this same set of components, although as you'll see throughout this chapter, things can get pretty complex quickly.

Still, if you understand the basic parts, when things do get more complex, you'll have a much easier time figuring out what's going on, and mapping more complex representations of the basic parts back to the functionality that those parts play.

Here are the parts of a typical function:

➤ **The fun keyword.** This identifies the function itself, to Kotlin and of course to developers writing code.

➤ **The function name.** This name is typically used in calling the function. As you'll soon see, though, many functions don't have names.

➤ **The argument list.** Functions can take in information via passed arguments, or parameters. You'll know you're looking at an argument list because the arguments are surrounded by parentheses. If there are no arguments, you'll still have an empty set of parentheses: ().

➤ **The return type of the function.** This is the only thing not shown in Figure 10.1. Functions can return values, and that's indicated by a colon (:) after the argument list and then the return type, like: fun copy(obj: Any) : Any.

➤ **The beginning of the function.** Functions typically open with an opening curly brace: {. This is the beginning of what the function actually does, also called the function's code block, which establishes the function's behavior.

➤ **The body of the function.** The body of the function exists between the beginning ({) and the end (}) of the function. There can be all sorts of code, or there can be nothing in the function body.

➤ **The end of the function.** This is the matching ending to the earlier beginning. Just as functions open with a curly brace, most close with one as well.

> **NOTE** *Notice qualifying words in the preceding list such as "most" and "typically." You're going to learn about these qualifiers, and the exceptions that require them, throughout this chapter.*

Function Arguments Also Have a Pattern

Functions are nothing new at this point. You've been calling them, with and without arguments, for 10 chapters now. Listing 10.2 is the `Person` class from Chapter 6, which has a number of functions.

LISTING 10.2: The Person class shown in Chapter 6

```kotlin
package org.wiley.kotlin.person

open class Person(_firstName: String, _lastName: String,
                  _height: Double = 0.0, _age: Int = 0) {

    var firstName: String = _firstName
    var lastName: String = _lastName
    var height: Double = _height
    var age: Int = _age

    var partner: Person? = null

    constructor(_firstName: String, _lastName: String,
                _partner: Person) :
            this(_firstName, _lastName) {
        partner = _partner
    }

    fun fullName(): String {
        return "$firstName $lastName"
    }

    fun hasPartner(): Boolean {
        return (partner != null)
    }

    override fun toString(): String {
        return fullName()
    }

    override fun hashCode(): Int {
        return (firstName.hashCode() * 28) + (lastName.hashCode() * 31)
    }

    override fun equals(other: Any?): Boolean {
        if (other is Person) {
            return (firstName == other.firstName) &&
                    (lastName == other.lastName)
```

continues

LISTING 10.2 *(continued)*

```
        } else {
            return false
        }
    }
}
```

You call these functions by using the function name, and if they're defined as part of a class (as is the case with `Person`), that function name comes after the class instance name:

```
val brian = Person("Brian", "Truesby", 68.2, 33)
println(brian.fullName())
```

Here, `fullName()` is called, and that particular function has no arguments, so there's nothing between the opening and closing parentheses.

If there is an argument, it's just inserted between the parentheses:

```
val brian = Person("Brian", "Truesby", 68.2, 33)
val rose = Person("Rose", "Bushnell", brian)

if (brian.equals(rose)) {
    println('These two people are the same.')
}
```

Here, `equals()` is called and the `rose` instance is passed in to the function. Here's the definition of that function, from Listing 10.2:

```
override fun equals(other: Any?): Boolean {
    // code
}
```

So function arguments appear as:

➤ Argument name

➤ Colon separator (:)

➤ Argument type

➤ Any modifiers, like the ? that indicates the parameter is optional

➤ Default values, if there are any

You can see the default values for `_height` and `_age` in the primary constructor to `Person`:

```
open class Person(_firstName: String, _lastName: String,
                  _height: Double = 0.0, _age: Int = 0) {
```

> **NOTE** *It's worth asking yourself, "Why are we revisiting this basic stuff?" It's a fair question, as you've been working with functions and arguments and default values for quite a while now. Still, it's important to get into the mechanics of these ideas for two reasons: first, you need to have a firm grasp of the syntax of any language you use that goes beyond what you've figured out implicitly; and second, this chapter will start to stretch these concepts, so ensuring you've got the basics down first will allow you to move into these more advanced uses much more quickly.*

Default Values in Constructors Are Inherited

Default values for constructors have some wrinkles when inheritance is introduced into the picture. The primary constructor for `Person` has default values declared like this:

```
open class Person(_firstName: String, _lastName: String,
                  _height: Double = 0.0, _age: Int = 0) {
```

If you add arguments to a constructor in a subclass of `Person`, you have to maintain those default values, redefine them, or ensure that you don't call the superclass version illegally.

For example, here's a `Child` subclass of `Person` that keeps the same default values for `_height` and `_age`:

```
class Child(_firstName: String, _lastName: String,
            _height: Double = 0.0, _age: Int = 0, _parents: Set<Person>) :
    Person(_firstName, _lastName, _height, _age) {
```

Here's the same subclass, but this time, the default `_age` is set to 12 instead of 0, as it is in `Person`:

```
class Child(_firstName: String, _lastName: String,
            _height: Double = 0.0, _age: Int = 12, _parents: Set<Person>) :
    Person(_firstName, _lastName, _height, _age) {
```

Now, you can legally remove the default value for `_height`:

```
class Child(_firstName: String, _lastName: String,
            _height: Double, _age: Int = 12, _parents: Set<Person>) :
```

However, the current code in the call to the superclass constructor is now illegal:

```
class Child(_firstName: String, _lastName: String,
            _height: Double, _age: Int = 0, _parents: Set<Person>) :
    Person(_firstName, _lastName, _height, _age) {
```

This is passing in `_height`, which may be undefined, and Kotlin won't allow that.

Default Values in Functions Are Inherited

Functions work slightly differently in inheritance, only because a primary constructor *must* call the superclass constructor, so that adds some requirements. Functions that are not constructors are a little broader in what they'll allow.

Here's a simple function with some default values:

```
fun add(num1: Int, num2: Int, num3: Int = 0, num4: Int = 0): Int {
    return num1 + num2 + num3 + num4
}
```

> **WARNING** *Sometimes good code makes for a poor example, and sometimes a good example makes for poor production code. This is an example of the latter. This is a terrible approach for building an* add() *function, but it does make default arguments and their usage very clear. Learn the concept, but don't try breaking out this* add() *function around your Kotlin peers; they likely won't appreciate its purpose!*

If you change this function to open, then it can be extended in a subclass. Listing 10.3 gives the source code for `Calculator`, based on work from Chapter 9 and this new `add()` function. Both `Calculator` and `add()` are marked as open, and available for subclassing.

LISTING 10.3: An open class with an open add() function

```
package org.wiley.kotlin.math

import kotlin.math.pow

open class Calculator {
    fun execute(input: Int, operation: Operation): Int {
        val returnVal : Int = when (operation) {
            is Operation.Add -> input + operation.value
            is Operation.Subtract -> input - operation.value
            is Operation.Multiply -> input * operation.value
            is Operation.Divide -> input / operation.value
            is Operation.Raise -> (input.toFloat().pow(operation.value
.toFloat())).toInt()
            else -> {
                throw Exception("Unhandled Operation: ${operation}")
            }
        }

        return returnVal;
    }

    open fun add(num1: Int, num2: Int, num3: Int = 0, num4: Int = 0): Int {
        return num1 + num2 + num3 + num4
    }
}
```

> **NOTE** *You'll also need the* `Operation` *class from Chapter 9 in your project to compile* `Calculator`.

If you now subclass `Calculator`—call it `Abacus`—you can override `add()`. Listing 10.4 shows this.

LISTING 10.4: Extending Calculator and overriding the add() function

```
package org.wiley.kotlin.math

class Abacus : Calculator() {

    override fun add(num1: Int, num2: Int, num3: Int, num4: Int): Int {
        return super.add(num1, num2, num3, num4)
    }
}
```

This overridden version of `add()` just calls the implementation of `add()` from its superclass, `Calculator`. But note that the default values of `add()` are left out:

```
override fun add(num1: Int, num2: Int, num3: Int, num4: Int): Int {
```

Default Values in Functions Cannot Be Overridden

In fact, you cannot specify those values even if you want. Update the function signature to look like this:

```
override fun add(num1: Int, num2: Int, num3: Int = 0, num4: Int = 0): Int {
```

Now compile, and you'll get this error:

```
Kotlin: An overriding function is not allowed to specify default values for its
parameters
```

So Kotlin won't allow you to specify those parameters, even if they're the same as the superclass definition. On top of that, the default values still apply. So you can call the `add()` function in `Abacus` like this:

```
var ab = Abacus()
println(ab.add(2, 1))
```

This actually can make your code a little less readable, because if you were to just glance at the code signature for `add()` in `Abacus`, you could miss that there are default values available. This is where a good IDE can become invaluable. Notice in Figure 10.2 that IntelliJ gives you the function signature, including the default values that are inherited from the `Calculator` base class.

Default Values Can Affect Calling Functions

Another wrinkle that default values for function arguments can have arises when you give an argument earlier in a list a default value, but then do *not* provide default values for later arguments. Update `add()` in `Calculator` to look like this:

```
open fun add(num1: Int, num2: Int = 0, num3: Int = 0, num4: Int): Int {
    return num1 + num2 + num3 + num4
}
```

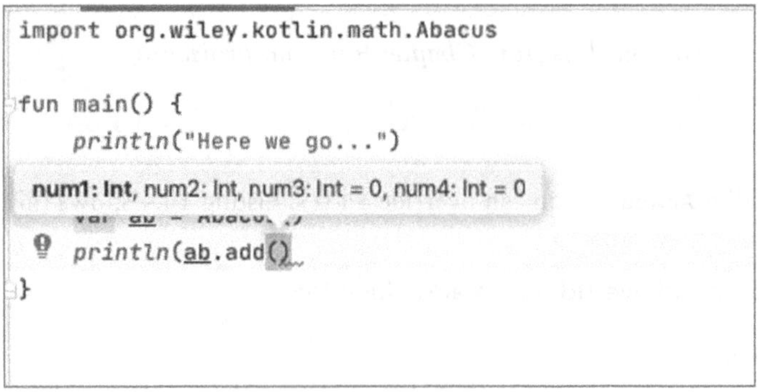

```
import org.wiley.kotlin.math.Abacus

fun main() {
    println("Here we go...")
    num1: Int, num2: Int, num3: Int = 0, num4: Int = 0
    var ab = Abacus()
    println(ab.add()
}
```

FIGURE 10.2: A good IDE gives you default values for a function, even from a base class.

Note that now num2 and num3 have default values but num4 does not.

> **WARNING** *You'll have to make this change to the version of* add() *in* Calcu-
> lator, *and not the overridden version in* Abacus. *The version in* Abacus *cannot
> change the default values, so the preceding code would not compile.*

Anytime you have default values for arguments that precede additional arguments without default values, some special rules come into play. If you do not name your arguments in the *calling* code—not the function code itself, but the code that *calls* that function—then no default values will be assumed. That means that this code will generate a compiler error:

```
println(ab.add(1, 2))
```

There is a combination of arguments that would make this call valid. If 1 was assigned to num1 and 2 to num4, then num2 would be given a default value of 0 and num3 would also be given a default value of 0. However, Kotlin is not willing to make this assumption.

Instead, absent any specific information to the contrary, Kotlin assumes your values are provided to the arguments in the function called, in order, starting with the first argument, and proceeding left to right. That means that 1 goes to num1 and 2 goes to num2—regardless of num2 and num3 having default values.

If you want to take advantage of those default values, you'll need to use named arguments, which you've seen before. So you could call add() like this:

```
println(ab.add(num1 = 1, num4 = 3))
```

This compiles, because now num2 and num3 both assume their default values. You can also skip the first argument name, because Kotlin already assumes that the first value is for num1:

```
println(ab.add(1, num4 = 3))
```

This is all made a lot easier with an IDE, as you can see in Figure 10.3. Most IDEs will do something similar and show the arguments accepted as well as their default values, if any.

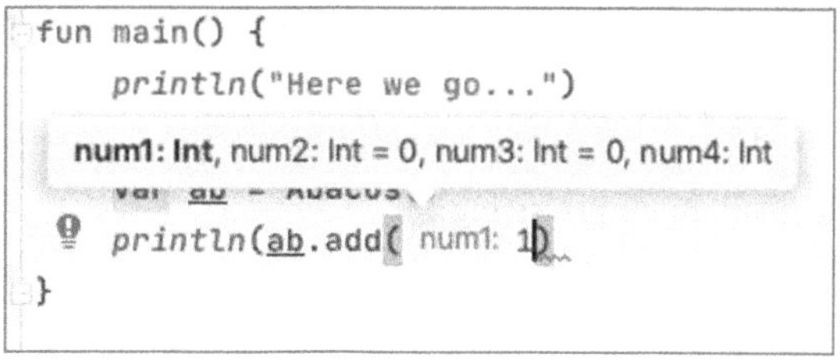

```
fun main() {
    println("Here we go...")
    num1: Int, num2: Int = 0, num3: Int = 0, num4: Int
    var ab = Abacus
    println(ab.add( num1: 1)
}
```

FIGURE 10.3: IDEs help with arguments, especially if you're naming arguments or working out of order.

Calling Functions Using Named Arguments Is Flexible

There's more available to you when calling functions than just using named arguments to work with default values. You can change the order of arguments as well:

```
println(ab.add(num4 = 3, num1 = 1, num2 = 8))
```

The preceding code does the same thing as this code:

```
println(ab.add(num1 = 1, num2 = 8, num4 = 3))
```

In fact, all the following are functionally identical (no pun intended!):

```
println(ab.add(num4 = 3, num1 = 1, num2 = 8))
println(ab.add(num1 = 1, num2 = 8, num4 = 3))
println(ab.add(num1 = 1, num2 = 8, num4 = 3, num3 = 0))
println(ab.add(1, num2 = 8, num4 = 3))
println(ab.add(1, 8, num4 = 3))
println(ab.add(1, 8, 0, 3))
```

In each case, a combination of named arguments, default values, and Kotlin's default ordering are used to get the same values into the same arguments.

Function Arguments Can't Be Null Unless You Say So

You've also seen how picky Kotlin is with regard to `null` values. You cannot pass `null` into a function as an argument value by default. Update `Abacus` to match Listing 10.5.

LISTING 10.5: Adding a function that prints a sum to Abacus

```
package org.wiley.kotlin.math

class Abacus : Calculator() {

    override fun add(num1: Int, num2: Int, num3: Int, num4: Int): Int {
        return super.add(num1, num2, num3, num4)
    }

    fun add_and_print(text: String, num1: Int, num2: Int, num3: Int, num4: Int)
    {
        println("${text} ${add(num1, num2, num3, num4)}")
    }
}
```

This new function uses the `add()` function already defined and prints out a sum with a `String` prefix. You can call this new function easily:

```
ab.add_and_print("The sum is:", 1, 8, 0, 3)
```

If you wanted to avoid the prefix string, you could send an empty string:

```
ab.add_and_print("", 1, 8, 0, 3)
```

> **NOTE** *This code makes the point about null values, but it also removes the ability to take advantage of default values that `add()` declares. To still allow those, you'd have to add the same default values that are defined in the `add()` function of `Calculator` (as they can't be restated in `Abacus`) to the function definition of `add_and_print()`.*
>
> *That's still a bad idea, because now if `Calculator`'s `add()` changes, you'll have to change `add_and_print()`, and that sort of undocumented tie between two functions—in two classes, no less—is a bad and fragile approach to code.*

However, you *cannot* call the function like this:

```
ab.add_and_print(null, 1, 8, 0, 3)
```

Kotlin won't accept null values, and gives you an error in compilation:

```
Kotlin: Null can not be a value of a non-null type String
```

You can avoid this by defining in the function that the `String` argument can be null:

```
fun add_and_print(text: String?, num1: Int, num2: Int, num3: Int, num4: Int) {
```

The `?` after the type of the argument for `text` tells Kotlin that you're prepared to take responsibility for `null` values.

Of course, getting code to compile and then running the code is two separate things. This code now compiles, but you're going to get silly output:

```
null 12
```

While `add_and_print()` allows null values, it really does nothing to actually handle them in a useful way. You'd want something more like this:

```
fun add_and_print(text: String?, num1: Int, num2: Int, num3: Int, num4: Int) {
    var outputString = StringBuilder();
    if ((text != null) && (!text.isEmpty())) {
        outputString.append(text).append(": ")
    }
    outputString.append(add(num1, num2, num3, num4))

    println(outputString)
}
```

In this version of `add_and_print()`, both `null` values and empty strings are handled intelligently. Now all three of these calls to the function behave in a sensible manner:

```
ab.add_and_print("The sum is", 1, 8, 0, 3)
ab.add_and_print("", 1, 8, 0, 3)
ab.add_and_print(null, 1, 8, 0, 3)
```

The first prefixes the sum and the second and third ignore a `String` prefix. Here's the output:

```
 The sum is: 12
12
12
```

FUNCTIONS FOLLOW FLEXIBLE RULES

All alliteration of that heading aside, everything up until this point is largely a review of things you've already done and understand. However, functions offer a bit more flexibility in their syntax and construction. They can appear to break the rules when, in fact, the rules are just a little more malleable than apparent when you first start writing functions.

This section covers that malleability and will help you bend functions even further to your own needs.

Functions Actually Return Unit by Default

If you don't want your function to return anything, you'd typically write it like this:

```
fun add_and_print(text: String?, num1: Int, num2: Int, num3: Int, num4: Int) {
```

There is nothing particularly exciting here, and you've been writing functions like this for a while now. But these functions actually *do* have a return value: `Unit`. `Unit` is a type you've never seen and is really more about language syntax than something you'd use in your code explicitly.

So, the preceding function actually looks like this once compiled:

```
fun add_and_print(text: String?, num1: Int, num2: Int, num3: Int, num4:
Int) : Unit {
```

`Unit` has no value except for `Unit`, which is odd, but reflects that this is really for Kotlin's own internal use. Additionally, unlike a function that returns a `String` or `Boolean` or another named type, a function that indicates it returns `Unit` doesn't *have* to explicitly return that type.

So if you look at the entire body of `add_and_print()`, it doesn't have an explicit `return` statement:

```
fun add_and_print(text: String?, num1: Int, num2: Int, num3: Int, num4: Int) {
    var outputString = StringBuilder();
    if ((text != null) && (!text.isEmpty())) {
        outputString.append(text).append(": ")
    }
    outputString.append(add(num1, num2, num3, num4))

    println(outputString)
}
```

However, you can add one:

```kotlin
fun add_and_print(text: String?, num1: Int, num2: Int, num3: Int, num4: Int) {
    var outputString = StringBuilder();
    if ((text != null) && (!text.isEmpty())) {
        outputString.append(text).append(": ")
    }
    outputString.append(add(num1, num2, num3, num4))

    println(outputString)

    return Unit
}
```

As odd as that last line looks, it's perfectly valid Kotlin. You can also leave off the Unit return type, and just add `return`:

```kotlin
fun add_and_print(text: String?, num1: Int, num2: Int, num3: Int, num4: Int) {
    var outputString = StringBuilder();
    if ((text != null) && (!text.isEmpty())) {
        outputString.append(text).append(": ")
    }
    outputString.append(add(num1, num2, num3, num4))

    println(outputString)

    return
}
```

Because this looks so odd and is potentially confusing, `return Unit` doesn't mean much to anyone unless you have a firm grasp of Kotlin internals—it's optional, and usually left out.

> **NOTE** *If you're a Java veteran, you might wonder why Kotlin doesn't support just returning* void, *which is what Java does in a comparable situation. This comes back to Kotlin's emphasis on strong typing. Kotlin always wants to return a type, and* void *isn't a type. So Kotlin uses* Unit *instead. This ensures that every function returns a type, all the time, and further reinforces that types are always present in Kotlin.*

Functions Can Be Single Expressions

Here's another look at the `add()` function that's been the subject of the earlier sections of this chapter. This is the version in `Calculator`:

```kotlin
open fun add(num1: Int, num2: Int = 0, num3: Int = 0, num4: Int): Int {
    return num1 + num2 + num3 + num4
}
```

In this function, there's a single line. Further, that single line is just one expression. In other words, if you squint a little bit and ignore the arguments for a moment, you could almost see something like this:

```kotlin
fun add() = num1 + num2 + num3 + num4
```

The entirety of add() boils down to just making a calculation. Put into Kotlin parlance, the function is a single expression.

Single-Expression Functions Don't Have Curly Braces

When you run across single-expression functions, you can skip the opening and closing curly braces, and assign the result of that expression to the function declaration. Listing 10.6 revisits Calculator with this notation for add().

LISTING 10.6: Changing add() to be a single expression assigned to the function

```
package org.wiley.kotlin.math

import kotlin.math.pow

open class Calculator {
    fun execute(input: Int, operation: Operation): Int {
        val returnVal : Int = when (operation) {
            is Operation.Add -> input + operation.value
            is Operation.Subtract -> input - operation.value
            is Operation.Multiply -> input * operation.value
            is Operation.Divide -> input / operation.value
            is Operation.Raise -> (input.toFloat().pow(operation.value
.toFloat())).toInt()
            else -> {
                throw Exception("Unhandled Operation: ${operation}")
            }
        }

        return returnVal;
    }

    open fun add(num1: Int, num2: Int = 0, num3: Int = 0, num4: Int): Int =
        num1 + num2 + num3 + num4
```

While this is a relatively simple change, it opens up a lot of interesting possibilities.

It's worth reviewing the original definitions of the beginning and end of a function:

➤ **The beginning of the function.** Functions typically open with an opening curly brace: {. This is the beginning of what the function actually does, also called the function's behavior.

➤ **The end of the function.** This is the matching ending to the earlier beginning. Just as most functions open with a curly brace, most close with one as well.

Notice that functions "typically open" with a curly brace. Now you're seeing that while that is true, they can also open with an equals sign (=). They also typically close with a closing curly brace, but only to match an open one, if present.

Single-Expression Functions Don't Use the return Keyword

Additionally, notice that in this expression format, there is no `return`. The result of the expression is returned. Further, that result must match the specified type, which is `Int` for `add()`.

Third, and most important for much of the rest of this chapter, you're now clearly seeing that expressions can stand in for all sorts of things, and that a function can actually *be* an expression. While it's been useful to think of a function as a declaration and a body, and that body is filled with executed statements, you can condense that a bit. A function is just an expression (maybe more than one), and at least part of the result of executing some or all of that expression.

Why does this matter? Keep reading, as this will quickly become a lot more dynamic. First, though, one last word on expression-based functions.

Single-Expression Functions Can Infer a Return Type

It's actually possible to streamline functions a little more when they evaluate to an expression. You can drop the return type, and Kotlin will figure out that return type from the type of the expression. So you can remove the `Int` return type from `add()`:

```
open fun add(num1: Int, num2: Int = 0, num3: Int = 0, num4: Int) =
    num1 + num2 + num3 + num4
```

Here, the expression is evaluated, which becomes an `Int`, and so `add()` is interpreted as returning an `Int`. So if you assigned the result to a `Float`, you'll have issues, as shown here:

```
var returnVal: Float
returnVal = ab.add(num4 = 3, num1 = 1, num2 = 8)
```

This code generates an error:

```
Kotlin: Type mismatch: inferred type is Int but Float was expected
```

You can change `returnVal` back to an `Int` and things will compile again.

Now, as useful as this inference is, it's not without downsides. As you've certainly seen many times over by now, Kotlin has type safety in its very makeup. Whether it's the refusal to accept null values unless explicitly declared with a ? in an argument list or the refusal to perform even basic type casting unless instructed, Kotlin doesn't like to infer types.

In this case, this inference can be problematic if you're not careful with your code that the type is inferred from. Take a look at Listing 10.7, which is the `Calculator` class from the last chapter revisited.

LISTING 10.7: Revisiting type inference in Calculator

```
package org.wiley.kotlin.math

import kotlin.math.pow

open class Calculator {
    fun execute(input: Int, operation: Operation): Int {
        val returnVal : Int = when (operation) {
```

```
            is Operation.Add -> input + operation.value
            is Operation.Subtract -> input - operation.value
            is Operation.Multiply -> input * operation.value
            is Operation.Divide -> input / operation.value
            is Operation.Raise -> (input.toFloat().pow(operation.value.
toFloat())).toInt()
            else -> {
                throw Exception("Unhandled Operation: ${operation}")
            }
        }
    }

    return returnVal;
}

open fun add(num1: Int, num2: Int = 0, num3: Int = 0, num4: Int) =
    num1 + num2 + num3 + num4
}
```

In particular, note that within `execute()`, there's only a single expression: the `when`. That means you can change this function to use the `=` notation instead of curly braces, like so:

```
fun execute(input: Int, operation: Operation): Int =
    when (operation) {
        is Operation.Add -> input + operation.value
        is Operation.Subtract -> input - operation.value
        is Operation.Multiply -> input * operation.value
        is Operation.Divide -> input / operation.value
        is Operation.Raise -> (input.toFloat().pow(operation.value
.toFloat())).toInt()
        else -> {
            throw Exception("Unhandled Operation: ${operation}")
        }
    }
```

> **NOTE** *If this change was confusing, just consider that the entire function was a single expression with the result being assigned to* returnVal, *and then* returnVal *was returned. That's a great indicator that you might have a single-expression function. By removing* returnVal *and simply making that assignment directly to the function itself, you've got yourself a working conversion.*

Now, you can go one step further, and remove the return type from the entire function (currently set as `Int`), and let Kotlin infer the return type:

```
fun execute(input: Int, operation: Operation) =
    when (operation) {
        is Operation.Add -> input + operation.value
        is Operation.Subtract -> input - operation.value
        is Operation.Multiply -> input * operation.value
        is Operation.Divide -> input / operation.value
        is Operation.Raise -> (input.toFloat().pow(operation.value
.toFloat())).toInt()
```

```
    else -> {
        throw Exception("Unhandled Operation: ${operation}")
    }
}
```

This will still compile and work without any issue.

Type Widening Results in the Widest Type Being Returned

But notice the case for handling `Operation.Raise` in the `execute()` function of the `Calcu-lator` class:

```
is Operation.Raise -> (input.toFloat().pow(operation.value.toFloat())).toInt()
```

When writing this case, it started out like this:

```
is Operation.Raise -> input.toFloat().pow(operation.value.toFloat())
```

But at the time of compilation, Kotlin threw an error. Because `execute()` declared that it would return an `Int` (remember, this was before the function was converted to use = instead of {, }, and `returnVal`), this case was rejected. It returns a `Float`, rather than an `Int`, and all the other cases returned `Int`s.

Now, though, change the code to match:

```
is Operation.Raise -> input.toFloat().pow(operation.value.toFloat())
```

Without `execute()` declaring that it returns an `Int`, no errors are generated at compile time. But look at what is now happening:

```
fun execute(input: Int, operation: Operation) =
    when (operation) {
        is Operation.Add -> input + operation.value
        is Operation.Subtract -> input - operation.value
        is Operation.Multiply -> input * operation.value
        is Operation.Divide -> input / operation.value
        is Operation.Raise -> (input.toFloat().pow(operation.value
.toFloat())).toInt()
        else -> {
            throw Exception("Unhandled Operation: ${operation}")
        }
    }
```

For `Operation.Add`, `Operation.Subtract`, `Operation.Multiply`, and `Operation.Divide`, the return value will be an `Int`, because an `Int` is passed in via `input`, and an `operation.value` will always be an `Int`. Operations on two `Int`s—even with division—will return an `Int`.

> **WARNING** *Although the focus in this section is on type inference in single-expression functions, don't let that last sentence slip by you without some awareness. Kotlin will return an* `Int` *if you divide one* `Int` *by another* `Int`, *which is frankly unintuitive and potentially problematic. If you divide an* `Int` *with a value of 5 by an* `Int` *with a value of 2, you will not get back a* `Float` *or* `Double` *with the value of 2.5, but rather another* `Int`, *with the value of 2. You get the quotient as the answer, with any remainder (decimal or otherwise) ignored.*

However, using the `math.pow()` function a `Float` is returned. It also will only operate on `Float` or `Double`, so you're going to have `Float` values (or `Double` values) all around.

This merely generated an error and caused you to convert back to an `Int` (with `toInt()`) when `execute()` was explicitly set to return an `Int`. But now, type inference is used. To see the potential issue, try this:

```
var calc = Calculator()
var result: Int
result = calc.execute(5, Operation.Add(5))
```

Compile this code and you're going to get a surprising error:

```
Kotlin: Type mismatch: inferred type is Any but Int was expected
```

So what's going on here? Well, Kotlin is forced to infer a type from the expression attached to `execute()`, and it *cannot* infer different types for different cases in the `when`. In other words, Kotlin cannot say "Well, if the case is for `Operation.Add` and `Operation.Subtract`, infer that an `Int` is returned, but if the case is for `Operation.Raise`, infer that a `Float` is returned."

Instead, Kotlin does a *type widening*. It has to come up with a type that is wide enough to support all possible return types. So for `execute()`, the possible return values are `Int` and `Float`. The narrowest type in Kotlin that supports *both* of those is `Any` (which is quite a wide type indeed).

If you want to avoid type widening and control the specific type returned, you'll have to avoid type inference. Specify the function return type, and then let the compiler take care of warning you when you don't return that type.

> **NOTE** *Before you continue, go ahead and change* `execute()` *in* `Calculator` *to again return an* `Int`. *You'll also need to take the result of the expression for the* `Operation.Raise` *case and convert it back to an* `Int` *using* `toInt()`. *That will allow your code to compile so you can keep following along.*

Functions Can Take Variable Numbers of Arguments

Remember the `add()` function defined earlier in `Calculator`, and then inherited and overridden in `Abacus`? Here it is for reference as defined in `Calculator`:

```
open fun add(num1: Int, num2: Int = 0, num3: Int = 0, num4: Int) =
    num1 + num2 + num3 + num4
```

> **NOTE** *This is actually a great example of several things you've been learning.* `add()` *uses a single expression and* = *to return a value, and the type of that value is inferred.*

This function works but is a bit silly. What if you want to add five numbers? Or six?

This is a case where you really want to support a variable number of arguments. You want to pass *some undefined number* of Ints into add() and let it sum them all up. Kotlin considers these as *variable arguments*. They are represented by the vararg keyword.

Create a new function called sum(), as shown in the complete listing for Calculator in Listing 10.8.

LISTING 10.8: Using varargs in a new function in Calculator

```
package org.wiley.kotlin.math

import kotlin.math.pow

open class Calculator {
    fun execute(input: Int, operation: Operation) : Int =
        when (operation) {
            is Operation.Add -> input + operation.value
            is Operation.Subtract -> input - operation.value
            is Operation.Multiply -> input * operation.value
            is Operation.Divide -> input / operation.value
            is Operation.Raise -> (input.toFloat().pow(operation.value
.toFloat())).toInt()
            else -> {
                throw Exception("Unhandled Operation: ${operation}")
            }
        }

    open fun add(num1: Int, num2: Int = 0, num3: Int = 0, num4: Int) =
            num1 + num2 + num3 + num4

    open fun sum(num1: Int, vararg numbers: Int): Int {
      var sum = num1
      for (num in numbers)
        sum += num
      return sum
    }
}
```

sum() takes two arguments now:

➤   num1, an Int

➤   numbers, a vararg where each type is an Int

There are two arguments because a sum should really always involve two numbers, and this makes that clear by definition.

> **NOTE** *A* vararg *can actually have 0 arguments, so it's still possible to call* sum() *and just pass in one number. You could expand* sum() *to take in* num1, num2, *and* numbers *if you wanted to ensure there were always two input values. That becomes a matter of preference at that point.*

A vararg Argument Can Be Treated Like an Array

`numbers` can be treated as an `Array`, which makes this bit of code work via iteration:

```
for (num in numbers)
        sum += num
```

You'll also get access to other `Array` methods, like `size()` and `get()` and `iterator()`.

You can call the `sum()` function now like this:

```
println(calc.sum(5, 2))
```

But you can also call it like this:

```
println(calc.sum(5, 2, 4, 6))
```

And this is legal, too:

```
println(calc.sum(5, 2, 4, 6, 8, 10, 12, 14, 18, 200))
```

However, the input argument is *not* an `Array`. In other words, this code is *not* legal:

```
println(calc.sum(10, IntArray(5) { 42 }))
```

Here, the second argument is a quick array created via `IntArray`, a helper function that creates arrays like a factory. But this will result in an error:

```
Kotlin: Type mismatch: inferred type is IntArray but Int was expected
```

So while the input `vararg` to the `sum()` function can be treated like an `Array` *within* that function, an `Array` cannot be passed to `sum()` as an actual argument.

In many ways, this is the reason that `vararg` exists: to avoid having to pass in an array of values. You could easily rewrite `sum()` like this:

```
open fun sum2(num1: Int, numbers: Array<Int>): Int {
    var sum = num1
    for (num in numbers)
        sum += num
    return sum
}
```

This works, but now you have to expose that `Array` in calling the function:

```
println(calc.sum2(10, Array<Int>(5) { 42 }))
```

This isn't necessarily better or worse than `sum()` with a `vararg`, but the calling syntax just isn't as simple and direct.

FUNCTIONS IN KOTLIN HAVE SCOPE

In addition to the rules that govern how you write functions, functions also have scope. There are three different scopes for a function:

- ➤ Local
- ➤ Member
- ➤ Extension

Local Functions Are Functions Inside Functions

A local function means that the function is local *to* something else, and in Kotlin and in the context of functions, this usually means a function is local to another function.

Here's a working (albeit rather silly) example:

```
open fun sum(num1: Int, vararg numbers: Int): Int {
    var sum = num1

    fun add(first: Int, second: Int) =
            first + second
    for (num in numbers)
        sum = add(sum, num)

    return sum
}
```

Here, the function `add()` is declared within the function `sum()`, and can only be used by the code in `sum()`. What's interesting about local functions is that they can access variables in the outer scope. In other words, they can access variables in the containing function. So, you could rewrite this as follows:

```
open fun sum(num1: Int, vararg numbers: Int): Int {
    var sum = num1

    fun add(toAdd: Int) {
        sum += toAdd
    }
    for (num in numbers)
        add(num)

    return sum
}
```

Here, `add()` is using `sum`, even though `sum` isn't declared or passed into `add()`. That's typical of local functions, as they have access to variables not passed into them directly.

> **NOTE** *The containing function of a local function is called the* closure. *Variables that are local to the closure are available to the local function as well as the closure itself.*

Member Functions Are Defined in a Class

You've been using member functions since you got started with Kotlin way back in Chapter 1. `Calculator`'s `execute()`, `add()`, and `sum()` functions are all member functions. However, `sum()` also has a local function, also called `add()`.

This demonstrates that outside a local function's closure, names can be duplicated. In other words, there is no conflict between the `add()` member function of `Calculator`, and the `add()` local function within `sum()`, which is in turn a member function.

Extension Functions Extend Existing Behavior without Inheritance

If you have a class like `Calculator`, and you want to extend or build upon that class's behavior, you would likely subclass `Calculator` and add a new function or override an existing function. This is exactly what the `Abacus` class did. However, this case was pretty simple because you had control of both the `Calculator` and `Abacus` classes. You were able to set `Calculator` as `open` and then subclass it.

In this case, it was pretty easy to subclass `Calculator`. The class itself was `open`, there wasn't a lot of abstract behavior that had to be defined, and there was no compelling reason *not* to subclass `Calculator` with `Abacus`.

It's not always this easy to extend or add to a class's behavior, though. Sometimes you may want to add behavior to a class that you *don't* have access to, and the class itself isn't `open`. Or the class that has the behavior you want to add to has a lot of other behavior that you don't want to extend, or even inherit.

In all of these cases, an extension function may allow you to add or override behavior *without* using inheritance and subclassing.

> **NOTE** *If you're having a hard time thinking of a case where an extension function is handy, consider that every time you use a library of code, you're likely using code that isn't open for extension and that you don't have the ability to subclass. That's* exactly *the use case for extension functions.*

Extend an Existing Closed Class Using Dot Notation

Suppose you've got the `Person` class shown in Listing 10.9. To make this a bit more realistic, this version removes the `open` keyword that allows `Person` to be subclassed. It is now in effect a class that can't be added to.

LISTING 10.9: A closed version of the Person class

```
package org.wiley.kotlin.person

class Person(_firstName: String, _lastName: String,
             _height: Double = 0.0, _age: Int = 0) {

    var firstName: String = _firstName
    var lastName: String = _lastName
    var height: Double = _height
    var age: Int = _age
```

continues

LISTING 10.9 *(continued)*

```
    var partner: Person? = null

    constructor(_firstName: String, _lastName: String,
            _partner: Person) :
        this(_firstName, _lastName) {
        partner = _partner
    }

    fun fullName(): String {
        return "$firstName $lastName"
    }

    fun hasPartner(): Boolean {
        return (partner != null)
    }

    override fun toString(): String {
        return fullName()
    }

    override fun hashCode(): Int {
        return (firstName.hashCode() * 28) + (lastName.hashCode() * 31)
    }

    override fun equals(other: Any?): Boolean {
        if (other is Person) {
            return (firstName == other.firstName) &&
                    (lastName == other.lastName)
        } else {
            return false
        }
    }
}
```

So perhaps you've downloaded this code in a library—perhaps even a library without source code. So all you have is the compiled version of this. (Listing 10.9 is shown above so you can actually get a compiled version.)

Now suppose you want to add a marry() function that takes in another Person, sets the two Person instances' partner property to each other, and ensure that hasPartner() for both returns true. That would be a piece of cake ... if you could update the source code for Person. Or if you could subclass Person. But in this example, you don't have either option; you have a Person class that you can't modify and that you can't extend.

To extend a class in this case, you add an extension function, and with dot notation, you can get limited access to that class—as if the extension function is within the extended class's scope. That's why it's a class of function in terms of scope. It's not a local function, it's not a member function, but it still has a specific scoping.

Listing 10.10 is a new test class called ExtensionFunctionApp. In addition to creating a few Person instances, it adds a helper function called printPartnerStatus() that makes it easy to do some quick printing and indicates whether a Person instance has a partner or not.

LISTING 10.10: A test program for using extension functions

```
import org.wiley.kotlin.person.Person

fun main() {
    val brian = Person("Brian", "Truesby", 68.2, 33)
    val rose = Person("Rose", "Bushnell")

    printPartnerStatus(brian)
    printPartnerStatus(rose)
}

fun printPartnerStatus(person: Person) {
    if (person.hasPartner()) {
        println("${person.fullName()} has a partner named ${person.partner?
.fullName()}")
    } else {
        println("${person.fullName()} is single")
    }
}
```

> **NOTE** *Most of this should be pretty self-explanatory. The only potential wrinkle is the* ? *after* person.partner *in the* printPartnerStatus() *helper function. That* ? *indicates that* partner *might be null, and if it is, to simply stop the operation.* null *is actually returned from the expression to indicate that a* null *value was found.*
>
> *Also note that even though this function called* hasPartner() *first, you still have to use the* ? *operator to safely call a function attached to a potentially nullable property (*partner*).*

Again, remember the (fictional) situation: you want to add a marry() function to Person, but don't have the ability to change Person. You can define a function using the class name (Person), a dot (.), and then the new function name.

Here's an example of how you could code marry():

```
fun Person.marry(spouse: Person) {
    this.partner = spouse
    spouse.partner = this
}
```

This is almost trivial in its simplicity, and it's one particularly cool feature of Kotlin.

this Gives You Access to the Extension Class

In addition to the dot notation for the function—following the class being extended—there's another addition: the this keyword. This is where the scoping of an extension function gets a little unusual. this in an extension function gives you access to the class being extended (usually called the *extension class*).

So `this.partner` references the `partner` property of the extension class, which will be the instance of `Person` that `marry()` is called on at runtime.

You can see this in action—and with some more context—by updating `ExtensionFunctionApp` to look like Listing 10.11.

LISTING 10.11: Adding an extension function to Person

```
import org.wiley.kotlin.person.Person

fun main() {
    val brian = Person("Brian", "Truesby", 68.2, 33)
    val rose = Person("Rose", "Bushnell")

    fun Person.marry(spouse: Person) {
        this.partner = spouse
        spouse.partner = this
    }
    printPartnerStatus(brian)
    printPartnerStatus(rose)
    brian.marry(rose)
    printPartnerStatus(brian)
    printPartnerStatus(rose)
}

fun printPartnerStatus(person: Person) {
    if (person.hasPartner()) {
        println("${person.fullName()} has a partner named ${person
.partner?.fullName()}")
    } else {
        println("${person.fullName()} is single")
    }
}
```

When a call is made to `brian.marry()`, `this` in `marry()` refers to the `brian Person` instance. If you called `rose.marry()`, then `this` would refer to the `rose` instance.

If you run this code, you'll see that Kotlin treats `marry()` just like any other function, with no distinction between it and functions that were coded directly into `Person`:

```
Brian Truesby is single
Rose Bushnell is single
Brian Truesby has a partner named Rose Bushnell
Rose Bushnell has a partner named Brian Truesby
```

FUNCTION LITERALS: LAMBDAS AND ANONYMOUS FUNCTIONS

Functions in Kotlin are typically declared. They have a name and are scoped to a class in most cases. This is a typical member function of the `Person` instance, and you can call it using its name, `fullName()`:

```
fun fullName(): String {
        return "$firstName $lastName"
}
```

And this is also declared, as an extension function. It can be called on a `Person` instance with its name, `marry()`:

```
fun Person.marry(spouse: Person) {
        this.partner = spouse
        spouse.partner = this
}
```

Nothing too special here.

Anonymous Functions Don't Have Names

But functions do not have to be declared like this. You can essentially drop a function in anywhere, even in the middle of other code. Create a new test class—you can call it `LambdaApp`—and add the code shown in Listing 10.12.

LISTING 10.12: Creating an anonymous function

```
fun main() {

    // Anonymous function
    fun() { println("Here we go!") }
}
```

This looks a bit odd because it's not declared with a name, but it also isn't reusable in any way. It's just a function sitting in the middle of another function (declared as `main()`). What's odder is that if you run this code, you will *not* get output. By putting the code inside curly braces ({ and }), you're telling Kotlin you're defining a function. Functions have to be run to execute, so since this function isn't, you'll get no output at all.

This is an example of an *anonymous function*, and it's just what it sounds like it is: a function that doesn't have a name. It looks like a function, it acts like a function, and Kotlin sees it as a function. It simply doesn't provide a name to be called.

You can also add parameters and return types, as shown in Listing 10.13.

LISTING 10.13: Adding parameters and a return value to an anonymous function

```
fun main() {

    // Anonymous function
    fun(input: String) : String {
      return "The value is ${input}"
    }
}
```

However, this is still pretty useless. In fact, most IDEs will even point this out, as Figure 10.4 shows: the anonymous function is unused.

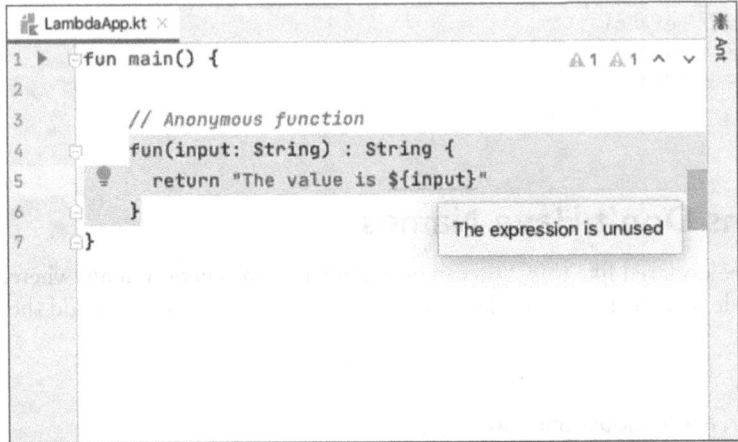

FIGURE 10.4: An unused anonymous function

You Can Assign a Function to a Variable

While some functions are anonymous and some are declared, all functions in Kotlin are considered *first class*. This means that they can actually be stored in variables. So update your test class to look like this:

```
var funVar = fun(input: String) : String {
  return "The value is ${input}"
}
```

Now you have a variable, funVar, that as its "value" has a function. It's important to note that this variable is not the *result* of running the function, but a reference to the function itself.

Try printing out the variable:

```
var funVar = fun(input: String) : String {
  return "The value is ${input}"
}
```

```
println(funVar)
```

You'll get something like this:

```
(kotlin.String) -> kotlin.String
```

To figure out what this is telling you, change your function definition:

```
var funVar = fun(input: Int) : String {
  return "The value is ${input}"
}
```

Now run the test program and see what's printed out:

```
(kotlin.Int) -> kotlin.String
```

Now you can see exactly what Kotlin is printing: a very brief representation of the function stored in funVar. It's a function that takes in an Int (or in the earlier version, a String), and returns (denoted by ->) a String.

Executable Code Makes for an "Executable" Variable

When you make the assignment to funVar, you're creating a named reference to that function—in this case, funVar. If you type in funVar and an open parenthesis ((), your IDE will actually suggest an argument—just as with any other function. You can see this in Figure 10.5.

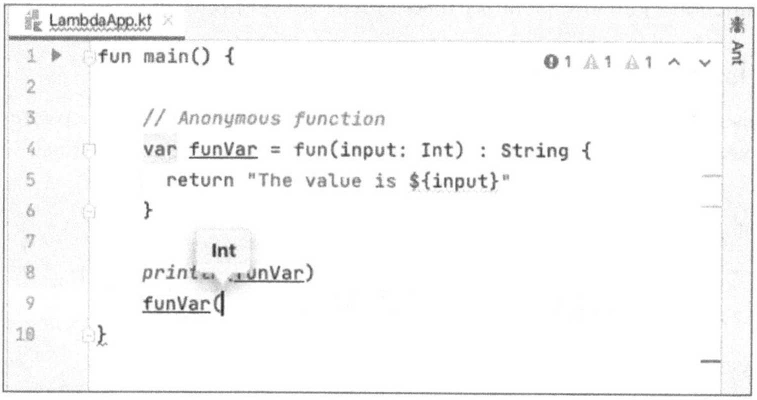

FIGURE 10.5: A function assigned to a variable is treated like the function itself.

You can drop in the argument like this:

```
var funVar = fun(input: Int) : String {
  return "The value is ${input}"
}

println(funVar)
funVar(2)
```

Now, run this code, and you'll still get the same output. That's because funVar returns a value, but nothing is being done with that value. What you can do to get that value into use is to use a function like println():

```
println(funVar(2))
```

Now you'll get some output:

```
println The value is 2
```

Higher-Order Functions Accept Functions as Arguments

You've come across something very important in this little line of code:

```
println(funVar(2))
```

`println()` is a higher-order function in Kotlin. Higher-order functions are functions that can take as arguments other functions. So here, you're actually passing a function (referenced by `funVar`) into the `println()` function. The function is treated like any other value, and is accepted by `println()` without an issue.

Note the definition of `println()` from Kotlin itself:

```
fun println(message: Any?)
```

Because this function takes `Any` (or nothing, which is allowed because of the `?`), it can accept function arguments. Consider if the signature were this:

```
fun println(message: String?)
```

This would no longer compile in cases where you passed a function argument into `println()`.

Any function that accepts a function as an argument is called a *higher-order function*.

The Result of a Function Is Not a Function

Before going further down this road, you need to make sure you're clear on the difference between a function and the result of executing a function. Take a look back at the `sum()` function defined in `Calculator`:

```
open fun sum(num1: Int, vararg numbers: Int): Int {
```

Now suppose you had another anonymous function defined and you passed that into `add()`:

```
calc.add( { 5 - 2 }, 3, 5, 3)
```

This will not compile. You'll get an error like this:

```
Kotlin: Type mismatch: inferred type is () -> Int but Int was expected
```

`add()` says it takes an `Int`, and while the result of executing the anonymous function is an `Int`, this code actually passes in the function itself. You can see this more clearly if you break up the statements:

```
var calc = Calculator()
var intFunction = { 5 - 2 }
calc.add(intFunction, 3, 5, 3)
```

`intFunction` is a function, not an `Int`. You'll get the same error, and it's even clearer why: you're giving `add()` the wrong argument type.

You would need to actually execute the function, which you can do with normal function notation: `()`. So here, `intFunction` is a reference to a function, and `intFunction()` will run that referenced function. You can use that to get this code working:

```
var calc = Calculator()
var intFunction = { 5 - 2 }
calc.add(intFunction(), 3, 5, 3)
```

Now the result of running the function is passed in, which works, as that result is an `Int`, exactly what `add()` expects.

Function Notation Focuses on Input and Output

If you go back to the error message from before you invoked `intFunction`, and just passed it in, you'll get a hint about how to declare functions as input arguments from that error message. Here's the error again:

```
Kotlin: Type mismatch: inferred type is () -> Int but Int was expected
```

You also saw something similar when you printed out an earlier variable that referenced a function:

```
(kotlin.Int) -> kotlin.String
```

In general, the notation is parentheses with any input argument types, an arrow (`->`), and a return type. So a function that took in an `Int` and a `String` and returns a `Boolean` would look something like this:

```
(kotlin.Int, kotlin.String) -> kotlin.Boolean
```

You can verify this by adding a statement like this to your test code:

```
println(fun(input: Int, input2: String) : Boolean =
    if (input > input2.length)
        true
    else
        false
)
```

Now suppose you actually want a function that takes in a function. For example, suppose you want a function that prints all the `String` entries in a list, but filters certain list items out. However, the function you're writing can take in *another* function that determines what is filtered out.

It helps to actually write the body of this function first. Suppose the list of `Strings` to print is called `input`, and the function to provide a filter is called `filterOut()`. That function will return `true` if the item should be filtered out (and not printed):

```
for (str in input)
  if (!filterOut(str))
      println(str)
```

Now that you know what you need, it's easier to define the function itself:

```
fun printList(input: List<String>, filterOut: [FUNCTION])
```

So while not legal Kotlin yet, this does take in the `input` list and a function named `filterOut`. For the actual type of `filterOut`, you can use a similar notation to what you've seen Kotlin give you already:

```
fun printList(input: List<String>, filterOut: (a: String) -> Boolean)
```

So `filterOut` takes in a function that has a `String` (referenced by `a`) as an argument and returns a `Boolean`. And *any* function that meets this criterion works.

So here's an example of a function you could pass in:

```
var filterA = fun(input: String) : Boolean =
        if (input.startsWith("A"))
            true
        else
            false
```

This function filters out any `String` that begins with a capital `A`. Here's another filter function:

```
var filterLong = fun(input: String) : Boolean =
        if (input.length > 6)
            true
        else
            false
```

Now the declaration is the same—both take in a `String` and return a `Boolean`—but the body of each filter function is different.

If you put all this together, you'll get something that looks like Listing 10.14.

LISTING 10.14: Testing out passing a function into a higher-order function

```
import org.wiley.kotlin.math.Calculator

fun main() {

    // Anonymous function
    var funVar = fun(input: Int) : String {
      return "The value is ${input}"
    }

    println(funVar)
    println(funVar(2))

    var calc = Calculator()
    var intFunction = { 5 - 2 }
    calc.add(intFunction(), 3, 5, 3)

    var filterA = fun(input: String) : Boolean =
            if (input.startsWith("A"))
                true
            else
                false
```

```
            var filterLong = fun(input: String) : Boolean =
                    if (input.length > 6)
                        true
                    else
                        false

            var strings = listOf<String>("Apple", "Carrot", "Horseradish",
                                        "Apricot", "Tomato", "Tangerine")

            println("Filtering out with filterA: ")
            printList(strings, filterA)

            println()
            println("Filtering out with filterLong: ")
            printList(strings, filterLong)
        }

        fun printList(input: List<String>, filterOut: (a: String) -> Boolean) {
          for (str in input)
            if (!filterOut(str))
                println(str)
        }
```

Here, `printList()` is a higher-order function that takes any function which accepts a `String` and outputs a `Boolean`. If you compile and run this code, you'll get different output from the two filters:

```
Filtering out with filterA:
Carrot
Horseradish
Tomato
Tangerine

Filtering out with filterLong:
Apple
Carrot
Tomato
```

> **NOTE** *There is some additional output omitted in the preceding example from earlier portions of the test code.*

You Can Define a Function Inline

While you can assign anonymous functions to variables, you can also just define the function as part of the function call itself. You saw an example of defining a function with curly braces earlier:

```
{ 5 - 2 }
```

> **NOTE** *This is actually something slightly different than an anonymous function. It's called a lambda expression and you'll be looking at that shortly in more detail.*

You can do the same thing when you call `printList`—except that you need a way to define an input argument. You can do that by supplying a variable name and type as the first part of the expression:

```
{ str: String -> false}
```

This is a function that takes in a `String` (referenced by `str`) and returns a `Boolean`, which in this case is always `false`. It's then not hard to extend this to actual logic. Here's a function declared in the same way that takes in a `String` and returns `true` if that `String` starts with a t or T:

```
{ str: String -> if (str.toUpperCase().startsWith("T")) true else false}
```

Lambda Expressions Are Functions with Less Syntax

What you're seeing here is not just an "inline function," but what Kotlin calls a *lambda expression*. This is just another way to write a function, and is even more broken down than an anonymous function.

Here's an anonymous function:

```
fun(str: String) : Boolean = if (str.toUpperCase().startsWith("T"))
    true
else
    false
```

Take out the `fun` keyword and parameter list, and it becomes even simpler. Now this is a lambda expression:

```
{ str: String -> if (str.toUpperCase().startsWith("T")) true else false}
```

> **NOTE** *It's largely just a convention that an anonymous function is called a function while a lambda expression is called an expression (and not a function). The only real syntactical difference is that more often than not, lambdas are shorter and even usually fit on a single line of a reasonable editor (think 80 characters or so).*
>
> *You will also occasionally see the term* lambda function. *This is really not technically correct, as a lambda expression can be used as a function but is not usually referred to as a function. This largely distinguishes lambda expressions from anonymous functions.*

Technically, there's more to a lambda expression. Here is the "fuller" version of the lambda expression just shown:

```
val filterT = { str: String -> if (str.toUpperCase().startsWith("T")) true
else false}
```

However, lambdas are often used without the `val` and just passed to another function inline, like this:

```
printList(strings, { str: String -> if (str.toUpperCase().startsWith("T")) true
else false})
```

Listing 10.15 shows the current complete version of LambdaApp, which uses lambda expressions as well as anonymous functions, all being passed to a higher-order function.

LISTING 10.15: A test of both anonymous and inline functions

```
import org.wiley.kotlin.math.Calculator

fun main() {

    // Anonymous function
    var funVar = fun(input: Int) : String {
      return "The value is ${input}"
    }

    println(funVar)
    println(funVar(2))

    var calc = Calculator()
    var intFunction = { 5 - 2 }
    calc.add(intFunction(), 3, 5, 3)

    var filterA = fun(input: String) : Boolean =
            if (input.startsWith("A"))
                true
            else
                false

    var filterLong = fun(input: String) : Boolean =
            if (input.length > 6)
                true
            else
                false

    var strings = listOf<String>("Apple", "Carrot", "Horseradish",
                        "Apricot", "Tomato", "Tangerine")

    println("Filtering out with filterA: ")
    printList(strings, filterA)

    println()
    println("Filtering out with filterLong: ")
    printList(strings, filterLong)

    val filterT = { str: String -> if (str.toUpperCase().startsWith("T")) true
else false}

    println()
    println("Filtering out with inline lambda: ")
    printList(strings, { str: String -> if (str.toUpperCase().startsWith("T")) true
else false})
}
```

continues

LISTING 10.15 *(continued)*

```
fun printList(input: List<String>, filterOut: (a: String) -> Boolean) {
    for (str in input)
        if (!filterOut(str))
            println(str)
}
```

You Can Omit Parameters Altogether

You've already seen that lambda expressions don't actually need parameters or a return type:

```
{ 5 - 2 }
```

Of course, this isn't particularly useful for much, so you won't see expressions like this often. You could theoretically use this notation if you wanted to always return the same value. Here's an example that filters out *all* input strings:

```
println()
println("Filtering out everything: ")
printList(strings, { true })
```

Note that even though `printList()` expects `(a: String) -> Boolean`, it will accept the lambda expression because the return value is the right type.

However, that exception doesn't apply if you mismatch parameters. So while it's okay to ignore parameters, you can't add an additional one. Try to add and compile this code:

```
printList(strings, { input1: String, input2: String -> true })
```

You'll get an error:

```
Kotlin: Expected one parameter of type String
```

This is because `printList()` sets an expectation that the function passed in will have a single `String` parameter:

```
fun printList(input: List<String>, filterOut: (a: String) -> Boolean) {
```

This actually makes a lot of sense. If a parameter is ignored, as is the case with a lambda expression that always returns true, it's possible to keep running the code. In essence, this:

```
{ true }
```

is equivalent to this:

```
{ str: String -> true }
```

The same is *not* true if you're trying to pass two parameters into a lambda expression that expects one parameter, or one parameter into a lambda expression that expects two parameters.

Lambda Expressions Use it for Single Parameters

Another idiom of lambdas is the use of a parameter called `it`. If you have a lambda that you want to accept a single parameter, you can omit the parameter and use `it` as a stand-in, even without declaring anything.

> **NOTE** *You may have noticed that the terms* lambda expression *and* lambda *are often used interchangeably. They are the same in terms of Kotlin use, so you'll often see* lambda *or* lambdas *as a shorter stand-in for* lambda expression *or* lambda expressions.

In the current definition of `filterOut`, the lambda takes in a single parameter. So you can actually omit that in the input lambda. Here's a version of a lambda that indicates the `parameter` it takes in:

```
{ str: String -> if (str.toUpperCase().startsWith("T")) true else false}
```

Here's an equivalent lambda using `it`:

```
{ if (it.toUpperCase().startsWith("T")) true else false}
```

Since this lambda only requires a single parameter, it's omitted. Then, the name `it` is used to represent that parameter.

Then this lambda is passed into a higher-order function; the higher-order function essentially controls the type of that parameter. So because `printList()` declares that it takes a lambda which accepts a `String`, that lambda expression basically can only take `String` inputs when used by `printList()`.

it Makes Lambdas Work More Smoothly

Now that you've seen how `it` works, you can get a better idea of why a lambda expression that has no declared input parameters can be passed in to a higher-order function that expects a function with a single input.

Consider this lambda expression:

```
{ str: String -> true }
```

In this expression, a single parameter is accepted, but then ignored in the actual expression. That's perfectly legal. Now recall that if you have a single parameter, you don't have to declare it. You can just use `it`:

```
{ if (it.length > 5) true else false }
```

Since you don't *have* to use input parameters, you can just as easily reduce all of this to:

```
{ true }
```

This is now a legal lambda expression for any higher-order function that accepts a function with a single argument.

Lambda Expressions Return the Last Execution Result

Take another look at some of the lambda expressions used for filtering so far:

```
{ if (it.toUpperCase().startsWith("T")) true else false}
```

But this can be made even simpler. A function or lambda can explicitly return a value. If there's not a `return`, then the expression is evaluated, and the result of evaluation is true. That's actually what is happening in the preceding code; the expression evaluates to either `true` or `false`, and that value is returned, even though there is not an explicit `return`.

Still, things can get even simpler. In lambda expressions with an `if`, all you may need is the `if` condition itself. If the condition is evaluated and comes back `true`, you can let that get passed up as the result of the lambda; if the condition evaluates to `false`, then that can be returned as well.

Here's how that might look for the same expression:

```
{ it.toUpperCase().startsWith("T") }
```

Through using `it` and just the condition from an `if`, this lambda expression gets quite concise, and is actually clearer to read.

Trailing Functions as Arguments to Other Functions

Just as `it` is idiomatic to lambdas in Kotlin, so is the ability to leave off parentheses on what is called a *trailing lambda*. Take a look at the definition of `printList()`:

```
fun printList(input: List<String>, filterOut: (a: String) -> Boolean) {
```

This function has as its *last* argument an input function. In this case, you will often see a different notation for calling this higher-order function. Currently, here's how `printList()` is called with a lambda expression input:

```
printList(strings, { it.toUpperCase().startsWith("T") })
```

However, you can include all but the trailing function in parentheses, and then leave the trailing lambda expression outside the parentheses:

```
printList(strings) { it.toUpperCase().startsWith("T") }
```

This probably looks a little odd, but you'll see this more often from experienced Kotlin programmers. It doesn't offer any particular advantage; it's just a notation change that you should be aware of when looking at other Kotlin code.

> **NOTE** *This truly is a matter of coding style and preference. You may find that you like this style, or you may find that you prefer to not use trailing lambdas, and keep everything in parentheses. There is no right or wrong, or even better or worse, here.*

LOTS OF FUNCTIONS, LOTS OF ROOM FOR PROBLEMS

With great functionality comes great room for poor programming. That's not quite a popular expression, but it could be. With anonymous functions, lambda expressions, higher-order functions, and everything you've learned prior to this chapter, you're seeing that almost any task in Kotlin can be accomplished in more than one way.

Additionally, with lambdas in particular, you can start to define functions quickly—often not even assigning them to variables—and pass them into other functions. You'll find greater flexibility than ever before as you use lambdas more and more.

However, there are two things to pay careful attention to as you get into the world of these varying approaches to defining functions:

➤ By using it, undeclared parameters, and the result of an expression as the return value, it can become increasingly unclear what your code does. Comment often and use shortcuts carefully to avoid your code becoming confusing and unreadable to others (and yourself when you revisit it days or weeks after initial writing).

➤ Be careful to not get sloppy with anonymous functions and lambdas. Sometimes it's easier to quickly type in a lambda, but it might be a better idea to create an actual function, declare it, and use it through its name. Functions really are about reuse.

Regardless of how often and how semantically you use these function tools, you *will* use them. In the next chapter, you're going to add some more Kotlin-specific idioms and patterns to your toolkit.

11

Speaking Idiomatic Kotlin

WHAT'S IN THIS CHAPTER?

➤ Using `let`, `run`, `with`, `apply`, and `also`

➤ Connecting code to scope

➤ Working with context objects

➤ Using scope functions for readability and style

SCOPE FUNCTIONS PROVIDE CONTEXT TO CODE

In Chapter 10, you saw that there are certain ways you can call lambdas that provide a sort of Kotlin-specific idiom. If a lambda is the last argument to a function—a trailing lambda—then you can omit parentheses around it.

This is an idiom, a term you've seen a few times. An idiom in language is a usage of words or terms, or in this case syntax, that is specific to a particular language, and often can't be interpreted literally. In English, you might say "get off my back!" and not mean anything related to an actual person on someone's back. But that idiom points to a thorough understanding of the language, and often distinguishes native or well-trained speakers from those who are not as familiar, or only have a more casual understanding, of the language.

In Kotlin, there are a number of idioms, and one of those is the trailing lambda. Several others are covered in this chapter—in particular, what Kotlin calls *scope functions*. A scope function lets you execute a block of code within a specific context, often that of a particular object instance.

> **NOTE** *To be clear, some of these functions and concepts, such as* let *and lambdas, appear in other languages. They are not unique to Kotlin, but they are often used in specific ways in Kotlin that reflect more advanced and experienced programmers.*
>
> *There's nothing that says you have to use idiomatic structures like trailing lambdas or scope functions in your code. However, if you take the time to learn them, you'll often get a better understanding of what the language is particularly good at, and therefore write better code that takes advantage of what Kotlin can really do.*

USE LET TO PROVIDE IMMEDIATE ACCESS TO AN INSTANCE

Listing 11.1 is an ordinary bit of code, something that should seem pretty easy to both write and understand at this point. It uses the `Calculator` and `Operation` classes from Chapters 9 and 10.

LISTING 11.1: Creating and using a Calculator instance

```
import org.wiley.kotlin.math.Calculator
import org.wiley.kotlin.math.Operation

fun main() {
    var calc = Calculator()
    var intFunction = { 5 - 2 }
    println(calc.add(intFunction(), 3, 5, 3))
    println(calc.execute(5, Operation.Add(4)))
    println(calc.execute(2, Operation.Add(4)))
}
```

There's nothing wrong with this code; however, it is a bit verbose, and of particular note, the `calc` variable really only exists to run a few functions. In other words, `calc` isn't itself valuable; it's just something that needs to be created so that the things that *are* of use can be done: running the `add()` and `execute()` functions.

Listing 11.2 is a reworking of this code using `let`.

LISTING 11.2: Using let to scope function calls

```
import org.wiley.kotlin.math.Calculator
import org.wiley.kotlin.math.Operation

fun main() {
    Calculator().let {
        var intFunction = { 5 - 2 }
        println(it.add(intFunction(), 3, 5, 3))
```

```
        println(it.execute(5, Operation.Add(4)))
        println(it.execute(2, Operation.Add(4)))
    }
}
```

This is a textbook example of a scope function, the `let` function. You can use `let` as a suffix on an object instance. In this case, an instance of `Calculator` is created:

```
Calculator()
```

Then, the dot notation is added, and `let` follows that. Then, a code block is introduced with an opening curly brace:

```
Calculator().let {
```

The context, or *scope*, of everything in that code block is the instance that was created. In other words, you could think of this code block as a function:

```
fun theFun(calc: Calculator) {
    var intFunction = { 5 - 2 }
    println(calc.add(intFunction(), 3, 5, 3))
    println(calc.execute(5, Operation.Add(4)))
    println(calc.execute(2, Operation.Add(4)))
}
```

Then, the new `Calculator` instance is created and passed *into* that function:

```
var calc = Calculator()
theFun(calc)
```

So now the function can operate on the `Calculator` instance.

Using `let`, you don't need to define a function, and you don't even need a variable to store the `Calculator` instance. It all happens through this scope function:

```
Calculator().let {
    var intFunction = { 5 - 2 }
    println(it.add(intFunction(), 3, 5, 3))
    println(it.execute(5, Operation.Add(4)))
    println(it.execute(2, Operation.Add(4)))
}
```

let Gives You it to Access an Instance

Once you create your basic `let` and associated code block, the instance that `let` is attached to is available via `it`:

```
println(it.add(intFunction(), 3, 5, 3))
println(it.execute(5, Operation.Add(4)))
println(it.execute(2, Operation.Add(4)))
```

This should feel familiar. It is exactly how lambda expressions worked when you didn't declare a single parameter:

```
{ if (it.length > 5) true else false }
```

The scope here is the instance created and passed into `let` via `it`. It is important to note that this instance *only exists* within this scope and therefore the code block. So, in this code:

```
Calculator().let {
    var intFunction = { 5 - 2 }
    println(it.add(intFunction(), 3, 5, 3))
    println(it.execute(5, Operation.Add(4)))
    println(it.execute(2, Operation.Add(4)))
}
```

there is no instance of `Calculator` available after the closing curly brace.

You can also rename `it` with your own name:

```
Calculator().let { calc ->
    var intFunction = { 5 - 2 }
    println(calc.add(intFunction(), 3, 5, 3))
    println(calc.execute(5, Operation.Add(4)))
    println(calc.execute(2, Operation.Add(4)))
}
```

Here, the `calc ->` notation indicates that the instance scoped into the code block is referenceable by `calc`. This new name overrides `it` and is not an additional reference. In other words, the following code will not compile:

```
Calculator().let { calc ->
    var intFunction = { 5 - 2 }
    println(it.add(intFunction(), 3, 5, 3))
    println(it.execute(5, Operation.Add(4)))
    println(it.execute(2, Operation.Add(4)))
}
```

Once you provide a custom name, `it` is no longer available. Only the instance via the custom name (in this case, `calc`) is.

The Scoped Code Blocks Are Actually Lambdas

You've already seen that the code blocks for a `let` (and you'll soon see for other scope functions as well) look a lot like lambda expressions. That's an accurate observation—and they *are* lambda expressions.

This also makes it clearer why `it` works as it does, as well as providing a customized name for the object instance that is passed into the lambda. These aren't new pieces of functionality, or specific to scope functions; they're just a part of how Kotlin handles lambda expressions.

Most scope functions—including `let`—also return the same value as the lambda expression would. So that's usually the result of evaluating the last line of the lambda.

Take the following code:

```
var result = Calculator().let {
    var intFunction = { 5 - 2 }
    println(it.add(intFunction(), 3, 5, 3))
    println(it.execute(5, Operation.Add(4)))
    println(it.execute(2, Operation.Add(4)))
```

```
    it.execute(5, Operation.Multiply(5))
}

println(result)
```

The variable `result` is assigned the value that's the execution of the last line of the `let` block—in this case:

```
it.execute(5, Operation.Multiply(5))
```

When you print that `result`, you get 25.

let and Other Scope functions Are Largely about Convenience

It's important to realize that scope functions like `let` and the ones you'll learn throughout the rest of this chapter are not introducing new functionality. There's nothing in the code you've seen so far in this chapter that is different than what you could have programmed without a `let`.

The main thing you gain with `let`, or any scoped function, is reduced "extra" code that doesn't add value. If you really don't need a `calc` variable, and instead just need to call `add()` once and `execute()` twice, then the code that uses `let` is *much* easier to read and leaves less clutter around your code.

However, if you are using `let` to reduce code clutter, you may want to take advantage of renaming the `it` reference, as you saw earlier.

You Can Chain Scoped Function Calls

Once you start to use scope functions, you'll start to use them *a lot*. A common example is chaining the result of one expression—inside a scoped function like `let`—and then using `let` again. Listing 11.3 provides an example.

LISTING 11.3: Chaining scope functions

```
import org.wiley.kotlin.math.Calculator
import org.wiley.kotlin.math.Operation

fun main() {
  var result = Calculator().let { calc ->
    var intFunction = { 5 - 2 }
    println(calc.add(intFunction(), 3, 5, 3))
    println(calc.execute(5, Operation.Add(4)))
    println(calc.execute(2, Operation.Add(4)))

    calc.execute(5, Operation.Multiply(5)).let {
        println("Inner result: ${calc.execute(it, Operation.Add(12))}")
    }
  }
}
```

This isn't too difficult to follow, especially because the outer `let` function uses a custom variable name for `it`: `calc`. `calc` is then used as it has been in previous examples.

Then, `calc.execute()` is called, but the result from that is handed off via an additional `let` lambda. This inner lambda does some printing, but uses the value from the outer call as input to the inner call:

```
calc.execute(5, Operation.Multiply(5)).let {
    println("Inner result: ${calc.execute(it, Operation.Add(12))}")
}
```

You could certainly rewrite this if you wanted:

```
var result = calc.execute(5, Operation.Multiply(5))
println("Inner result: ${calc.execute(result, Operation.Add(12))}")
```

There's not any benefit in terms of functionality here. However, as in previous cases, you save a variable you don't really need, and as long as the person reading the code understands `let` and scope functions, it's a more obvious flow.

It also re-emphasizes the point that scope functions are largely about convenience. While there is some small memory cost for storing an additional variable that might not be needed, the main thing you gain is legibility and a greater degree of connection between different parts of your code.

An Outer it "Hides" an Inner it

It is important to note that the previous code works because it revolves around the use of a custom name for the outer `it`:

```
var result = Calculator().let { calc ->
    var intFunction = { 5 - 2 }
    println(calc.add(intFunction(), 3, 5, 3))
    println(calc.execute(5, Operation.Add(4)))
    println(calc.execute(2, Operation.Add(4)))

    calc.execute(5, Operation.Multiply(5)).let {
        println("Inner result: ${calc.execute(it, Operation.Add(12))}")
    }
}
```

If you move the `calc ->` bit at the beginning of the lambda, things start to go awry. Remove that and replace the `calc` references with `it` throughout:

```
var result = Calculator().let {
    var intFunction = { 5 - 2 }
    println(it.add(intFunction(), 3, 5, 3))
    println(it.execute(5, Operation.Add(4)))
    println(it.execute(2, Operation.Add(4)))

    it.execute(5, Operation.Multiply(5)).let {
        println("Inner result: ${it.execute(it, Operation.Add(12))}")
    }
}
```

You'll get a compilation error if you try to run this code:

```
Kotlin: Unresolved reference: execute
```

This is referencing the final `execute()`, within the inner `let`:

```
println("Inner result: ${it.execute(it, Operation.Add(12))}")
```

The problem is that now `it` is ambiguous in this innermost lambda. It could refer to the outer `let` or the inner `let`. The outer `let` uses `it` to reference an instance of `Calculator`, and the inner `let` uses `it` to reference a number, the result of an `execute()` call (on the outer `Calculator` instance).

> **NOTE** *Take your time working through this section. It can get confusing when you have nested scope functions and you're working through the scoped variables. You may even want to print out the code, or make notes in the text, to ensure you can connect each `it` reference back to the correct `let` and instance that `it` connects with.*

Most IDEs will give you some feedback when you're potentially creating a conflict. Figure 11.1 shows you what IntelliJ will warn.

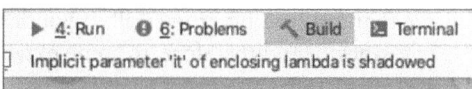

FIGURE 11.1: Good IDEs warn you of conflicts in scope function variables.

The message really gives a much clearer idea of the actual problem: "Implicit parameter 'it' of enclosing lambda is shadowed." This is in the context of the inner `let` and indicates that the `it` for that lambda is "shadowed" (or obscured) by the outer `it` reference.

Unfortunately, this warning can come and go in a flash, so you'll often end up relying on the compiler (which gives a slightly less clear message) to track down any problems. As a best practice, if you are going to have an inner scope function within an outer scope function, consider custom variable names to make scope very clear.

Chaining Scope Functions and Nesting Scope Functions Are Not the Same

In the code in Listing 11.3, there were two different things going on:

➤ An initial `let` scope function that had within it a second `let` scope function. This is an example of one scope function *nested* within another.

➤ A function call (`execute()`) where the output value was passed into a scope function. This is an example of chaining a scope function to another function.

It's important to understand the distinction between these two uses, from both a code-syntax (lexical) point of view and a usage point of view.

Nesting Scope Functions Requires Care in Naming

When you nest scope functions, you run the risk of the issue that was mentioned: the outer scope function's `it` reference will obscure or shadow the inner scope function's `it` reference. To avoid this, you'll need to use a custom reference variable in at least one of the scope functions, as you've seen when it was renamed to `calc` in the first `let` function earlier.

Some developers will go a little further, and as a rule, try to *always* provide custom names for lambda expressions. Here's that same code you saw earlier, but with `result` used for the inner lambda:

```
var result = Calculator().let { calc ->
    var intFunction = { 5 - 2 }
    println(calc.add(intFunction(), 3, 5, 3))
    println(calc.execute(5, Operation.Add(4)))
    println(calc.execute(2, Operation.Add(4)))

    calc.execute(5, Operation.Multiply(5)).let { result ->
        println("Inner result: ${calc.execute(result, Operation.Add(12))}")
    }
}
```

This isn't a bad idea, and really does make your code quite clear. However, it turns out to be a practice that can be a bit of overkill. If you have a very short lambda, you actually remove a lot of its "shortness" (and therefore, to some extent, its value) with this:

```
var finalVal = calc.execute(5, Operation.Multiply(5)).let { result ->
calc.execute(result, Operation.Add(12) }
```

> **NOTE** *As a bit of a case in point, the line of code in the example doesn't even fit in a single line of the book!*

In these cases—especially where you're doing a single, very clear thing with your lambda—using `it` is preferable:

```
var finalVal = calc.execute(5, Operation.Multiply(5)).let { calc.execute(it,
Operation.Add(12) }
```

It's much shorter and just as clear.

Chaining Scope Functions Is Simpler and Cleaner

It is quite a bit easier to simply chain scope functions. This is done through remembering a key aspect of any lambda expression and scope function: the value returned is the result of the final expression executed. Take this code:

```
var bigResult = Calculator().let { calc ->
    calc.execute(8, Operation.Add(4)).let {
        calc.execute(12, Operation.Multiply(it))
    }.let {
        calc.execute(it, Operation.Divide(18))
    }.let {
        calc.execute(16, Operation.Subtract(it))
    }.let {
        calc.execute(it, Operation.Raise(2))
    }
}

println("Big result is ${bigResult}")
```

Here, multiple expressions are chained together. Each expression is the last (and only) expression in each lambda handed to a `let`. So, the result of evaluating each is returned, and then passed to the next `let`.

You could also condense this further:

```
calc.execute(8, Operation.Add(4)).let { calc.execute(12, Operation.Multiply(it))
    }.let { calc.execute(it, Operation.Divide(18))
    }.let { calc.execute(16, Operation.Subtract(it))
    }.let { calc.execute(it, Operation.Raise(2))
    }
```

And you could condense this even further still, potentially onto one very long line.

The goal here isn't to write the shortest version, or the longest line of code. The goal is to avoid having variable after variable scattered around, each only being used once:

```
var calculator = Calculator()
var one = calc.execute(8, Operation.Add(4))
var two = calc.execute(12, Operation.Multiply(one))
var three = calc.execute(two, Operation.Divide(18))
var four = calc.execute(16, Operation.Subtract(three))
var five = calc.execute(four, Operation.Raise(2))
```

There is not a single value here other than the result (stored in `five`) that needs to hang around. Even `calculator` is really not needed after all the calls to `execute()` complete.

That's the core value proposition of scope functions: to make your code more convenient, organized, and easy to read.

Prefer Chaining over Nesting

It's really important to get comfortable with the idea of using chaining versus nesting, and in most cases, using chaining *over* nesting. You could write the previous code as a nested set of expressions:

```
var nestedResult = Calculator().let { calc ->
    calc.execute(8, Operation.Add(4)).let {
        calc.execute(12, Operation.Multiply(it)).let {
            calc.execute(it, Operation.Divide(18)).let {
                calc.execute(16, Operation.Subtract(it)).let {
                    calc.execute(it, Operation.Raise(2))
                }
            }
        }
    }
}

println("Nested result is ${bigResult}")
```

There are a number of obvious problems here, though. None of them will stop you from compiling your code, but they are all worth examining.

First, this code is much less readable, as you end up with all those closing brackets to match up. You'll have to make sure that each statement is properly indented, or it becomes a mess to read, as well.

> **NOTE** *IDEs don't particularly like this nested form, either. I kept having to remove a closing bracket and move it to even write this code in the first place. It's almost as if the IDE knew that there was a better approach!*

A second problem is that the usage of it can become hairy. While it's relatively clear that it applies to the current lambda, what it references is actually changing as you go down further into the nesting. That's potentially confusing.

What you might have noticed, though, is that this case of nesting does *not* result in the shadowing of an it reference. When you immediately attach the let (or other scope function) to the last lambda expression, you can use it as shown without concern.

Many Chained Functions Start with a Nested Function

There's another common pattern that shows up in this example:

```
var bigResult = Calculator().let { calc ->
    calc.execute(8, Operation.Add(4)).let {
        calc.execute(12, Operation.Multiply(it))
    }.let {
        calc.execute(it, Operation.Divide(18))
    }.let {
        calc.execute(16, Operation.Subtract(it))
    }.let {
        calc.execute(it, Operation.Raise(2))
    }
}

println("Big result is ${bigResult}")
```

It's the idea of a single nesting with lots of chained scope functions within that nesting. The initial let call on an instance of Calculator has within it a lot of nested scope functions, but each of those are chained together.

This turns out to be common, especially once you start to think of scope functions as ways to avoid keeping variables around that are only needed for a single purpose. You'll often create an initial instance that you can then operate on or with over and over, chaining results.

You Can Scope Functions to Non-Null Results

Another important use of let and scope functions is to address situations where an instance or variable might be null.

> **NOTE** *You're probably noticing that much of this chapter talks about let and scope functions in the same breath. While not all scope functions act identically to let, most follow the same basic rules. This section gets you comfortable with those rules, and the following sections detail the specific scope functions beyond let, and how each is similar to and varies from let.*

Listing 11.4 is a new program. Enter it and call it `NonNullApp` if you need a name.

LISTING 11.4: A simple program using code and nullable values

```
fun main() {
    var someString: String? = getRandomString(10)
    println(someString)
}

fun getRandomString(len: Int) : String? =
    when {
        len < 0  -> null
        len == 0 -> ""
        len > 0  -> (1..len).map { "ABCDEFGHIJKLMNOPQRSTUVWXYZabcdefghijklmnopqrstu
vwxyz".random() }.joinToString("")
        else -> null
    }
```

This code does two important things for the purpose of understanding scope functions in the context of nullable variables:

➤ Defines a nullable `String` variable, `someString`.

➤ Defines a function, `getRandomString()`, that creates a random `String` of a given length, and can potentially return `null`.

What's critical is understanding that the only reason that `println()` will accept `someString` as an input is because `println()` will accept `null` values. (It accepts `Any?` as a parameter.)

But more often than not, you won't want to allow `null` values into your functions, for all the type safety and related reasons you've been learning about throughout this entire book. So here's a very silly sample function that takes in a `String` and does some basic formatting to that `String`:

```
fun formatString(str: String) : String =
    "\n>>> ${str} <<<\n"
```

You can imagine more realistic things this function might do, but the key is that—like most functions—it accepts a typed input parameter that is *not* nullable. To see the effect of this, try to pass in `someString`, as shown in Listing 11.5.

LISTING 11.5: A non-compilable program with null problems that need to be fixed

```
fun main() {
    var someString: String? = getRandomString(10)
    println(someString)

    println(formatString(someString))
}

fun getRandomString(len: Int) : String? =
    when {
```

continues

LISTING 11.5 *(continued)*

```
        len < 0  -> null
        len == 0 -> ""
        len > 0  -> (1..len).map { "ABCDEFGHIJKLMNOPQRSTUVWXYZabcdefghijklmnopqrstu
vwxyz".random() }.joinToString("")
        else -> null
    }

fun formatString(str: String) : String =
        "\n>>> ${str} <<<\n"
```

Try to compile this, and you'll get an error:

```
Kotlin: Type mismatch: inferred type is String? but String was expected
```

This is a problem you've seen before. `formatString()` does not accept `String?`, but `String`. In other words, it will take a `String`, but that `String` *must* be non-null, and the Kotlin compiler is ensuring that is the case.

Accepting null Values Isn't a Great Idea

One easy fix here would be to simply change `formatString()` to have this signature:

```
fun formatString(str: String?) : String =
        "\n>>> ${str} <<<\n"
```

This is legal and will take care of the compilation error. With this change, `formatString()` will accept `someString`, because `someString` *might* be null, and that's okay as `formatString()` takes in a `String?`.

However, this is really a bad practice. One of the strongest features of Kotlin is its strong typing and compile-time type safety, and every time you intentionally code something that accepts nullable values, you lessen these strong typing and type safety abilities.

While you'll never be able to completely remove the `?` operator or functions and variables that might interact with `null`, you should try to reduce the occurrences of this.

Scope Functions Give You Null Options

So what do you do? You want to keep type safety and avoid lots of nullable values, but sometimes, you do have a situation where you *might* have a null and you need to deal with that. Thankfully, scope functions are perfect for this.

You can add a scope function like `let` onto a variable that might be `null` *after* you add a `?` operator. If the variable is `null`, the scope function is ignored. If it's not, the scope function executes. Listing 11.6 shows this in action.

LISTING 11.6: Using a scope function to work around potentially null values

```
fun main() {
    var someString: String? = getRandomString(10)
    println(someString)
```

```
        someString?.let { println(formatString(someString)) }
    }

    fun getRandomString(len: Int) : String? =
        when {
            len < 0  -> null
            len == 0 -> ""
            len > 0  -> (1..len).map { "ABCDEFGHIJKLMNOPQRSTUVWXYZabcdefghijklmnopqrstu
    vwxyz".random() }.joinToString("")
            else -> null
        }

    fun formatString(str: String) : String =
        "\n>>> ${str} <<<\n"
```

This code will now happily compile and run. Here's the key line:

```
    someString?.let { println(formatString(someString)) }
```

The output is what you'd expect:

```
    qbttNxelxV

    >>> qbttNxelxV <<<
```

The value of someString is first printed (showing it's not null), and then printed with formatting via formatString().

But now change the initialization of someString:

```
    var someString: String? = getRandomString(-2)
```

If you take a look at getRandomString(), you'll see that a negative input parameter will result in null being returned:

```
    fun getRandomString(len: Int) : String? =
        when {
            len < 0  -> null
            len == 0 -> ""
            len > 0  -> (1..len).map { "ABCDEFGHIJKLMNOPQRSTUVWXYZabcdefghijklmnopqrstu
    vwxyz".random() }.joinToString("")
            else -> null
        }
```

So someString is now null. This first line will work:

```
    println(someString)
```

That's because println() accepts Any?. But it is the following line that is interesting:

```
    someString?.let { println(formatString(someString)) }
```

someString? indicates to Kotlin that the variable (someString) might be null. If it is, execution stops. If someString is *not* null, then the println() with formatString() is executed.

In this situation, with someString null, you would *not* expect that scope function to run. And, it doesn't! There's no additional output:

```
    null
```

You see `null`, from the first `println(someString)`, and then there is no additional output. This verifies that the second `println()` which uses `formatString()` doesn't execute. Anytime you have a variable that may be `null`, you can use a scope function to specify activity on that variable *without* risking compiler errors.

Scope Functions Work on Other Functions . . . in Very Particular Ways

So now you know that taking a `null` variable and applying a lambda to it via a scope function works, and you can even avoid execution if a `null` is involved:

```
someString?.let { println(formatString(someString)) }
```

But remember, much of what scope functions give you is convenience and clarity of code. They remove variables you don't need from cluttering up your code, as well as memory.

In the previous example, though, there are still some wasted variables. Here's the fuller section:

```
var someString: String? = getRandomString(-2)
someString?.let { println(formatString(someString)) }
```

Although using the `let` is nice, `someString` serves no purpose other than to hold the result from calling `getRandomString()`. That's exactly the sort of thing that scope functions are intended to clean up.

You can take things further! You can apply a `let` (or any scope function) to a function, and if that original function returns `null`, the scope function won't run. Take a look at Listing 11.7, which demonstrates this idea after the code you've already written.

LISTING 11.7: Applying a scope function to a function that might return null

```
fun main() {
    // var someString: String? = getRandomString(10)
    var someString: String? = getRandomString(-2)
    println(someString)

    someString?.let { println(formatString(someString)) }

    getRandomString(12).let {
        println(formatString(it))
    }
}

fun getRandomString(len: Int) : String? =
    when {
        len < 0  -> null
        len == 0 -> ""
        len > 0  -> (1..len).map {
"ABCDEFGHIJKLMNOPQRSTUVWXYZabcdefghijklmnopqrstuvwxyz".random() }.joinToString("")
        else -> null
    }

fun formatString(str: String) : String =
    "\n>>> ${str} <<<\n"
```

This code pulls together several things you've already learned:

1. Rather than assigning `getRandomString()`'s result to a variable that will only be used once, it is called directly.

2. `getRandomString()` could return `null`, so a scope function is applied.

3. The lambda passed to the scope function uses the `it` reference to operate upon what comes into the function.

Now, there's no need for `someString` at all. It's only in Listing 11.7 for the previous example.

However, there's still a problem: this code won't actually compile. You'll get an error:

```
Kotlin: Type mismatch: inferred type is String? but String was expected
```

This is triggered by this statement:

```
println(formatString(it))
```

While `getRandomString()` is handled by the `let`—even if that function returns `null`—you've run into a bit of a quirk here. While the `let` on `getRandomString()` will only run if it's non-`null`, the function's return value is still `String?`. That means that from a very strict perspective, the `it` reference in the lambda is to a `String?`, not a `String`.

This presents a problem, because `formatString()` accepts a `String`. Remember, the whole point of this exercise was to *avoid* changing `formatString()` to take in a `String?` (accepting a `String` or a `null` value).

To really dig into what's going on, change your code to this:

```
getRandomString(-2).let {
    println("In here")
}
```

By passing in a negative value to `getRandomString()`, you'll get a `null` as the return value:

```
fun getRandomString(len: Int) : String? =
    when {
        len < 0  -> null
        len == 0 -> ""
        len > 0  -> (1..len).map { "ABCDEFGHIJKLMNOPQRSTUVWXYZabcdefghijklmnopqrstu
vwxyz".random() }.joinToString("")
        else -> null
    }
```

Now, you're passing that `null` to `let`. So the `let` code should *not* execute, right? You should compile the code and see. You'll get this surprising output:

```
In here
```

So, what in the world is going on? The problem is that a key part of syntax is missing: `?.let` and other scope functions don't automatically only execute on a non-`null` value. That functionality—or conditional execution, if you will—only happens if you explicitly indicate that the input might be `null`. This goes beyond the method declaration for `getRandomString()`, which clearly indicates it might return a `null`:

```
fun getRandomString(len: Int) : String? =
```

Instead, the `?` operator must be applied. Add the `?` to the end of the function call. It's actually the `?` that is telling Kotlin to only conditionally execute the lambda, not the scope function itself:

```
getRandomString(-2)?.let {
    println("In here")
}
```

This is a very important distinction. Scope functions can be set to only execute if the input is non-`null`, but the trigger for this is the `?`, *not* simply the presence of a scope function.

If you compile and execute this code, you'll see no output from the `println("In here")` statement, which is what you'd expect. Additionally, as you can see in Figure 11.2, a good IDE will also indicate that now the `it` reference is to a `String`, not a `String?`.

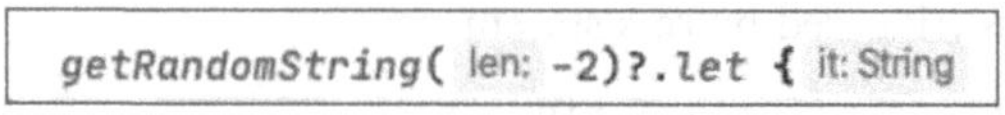

```
getRandomString( len: -2)?.let { it: String
```

FIGURE 11.2: Distinguishing between a String and a String?

Once you add the `?` after the function call, you can return the code to its original state, as shown in Listing 11.8.

LISTING 11.8: Using the ? operator to apply a scope function

```kotlin
fun main() {
    // var someString: String? = getRandomString(10)
    var someString: String? = getRandomString(-2)
    println(someString)

    someString?.let { println(formatString(someString)) }

    getRandomString(-2)?.let {
        println(formatString(it))
    }
}

fun getRandomString(len: Int) : String? =
    when {
        len < 0  -> null
        len == 0 -> ""
        len > 0  -> (1..len).map {
"ABCDEFGHIJKLMNOPQRSTUVWXYZabcdefghijklmnopqrstuvwxyz".random() }.joinToString("")
        else -> null
    }

fun formatString(str: String) : String =
    "\n>>> ${str} <<<\n"
```

With this change, you'll get no output if `getRandomString()` is null, and correct output if there's a `String` return value to format.

WITH IS A SCOPE FUNCTION FOR PROCESSING AN INSTANCE

Before digging too much into the specifics of another scope function, `with`, it's important that you recognize that all scope functions are very similar in use. While there are some syntactical differences, you'll quickly see that almost everything you learned about `let` applies to `with`.

You can take an object and operate upon it using `with`. Here's a basic example, again using an instance of `Calculator`:

```
with(Calculator()) {
    println(execute(8, Operation.Add(4)))
    println(execute(12, Operation.Multiply(12)))
}
```

You can quickly see what this does just by looking at the code.

1. First, the scope function is provided: `with()`.

2. A context object is provided to `with`. That context object, an instance of `Calculator` created through `Calculator()`, becomes the reference throughout the lambda.

3. Any operations within the `with` lambda are assumed to operate upon the context object.

Because the context object is an instance of `Calculator`, all the methods of that instance can be run: `execute()`, `add()`, and `sum()`.

> **WARNING** *This code sample and text assumes that you've kept* Calculator *current as you've gone through the various chapters. If not, you'll want to download the code samples so you can follow along and have available all the functions referenced throughout this and the following sections.*

with Uses this as Its Object Reference

One of the big differences between `let` and `with` is how the context object is referenced. In `let`, you used `it`, as is common in lambda expressions:

```
getRandomString(-2)?.let {
    println(formatString(it))
}
```

`with` uses `this` instead. Knowing that, you could rewrite the `with` block to be more explicit:

```
with(Calculator()) {
    println(this.execute(8, Operation.Add(4)))
    println(this.execute(12, Operation.Multiply(12)))
}
```

However, this is a bit clunky, and moves away from the point of this chapter: writing idiomatic Kotlin. Since anytime you want to invoke a function attached to the current `this` object, you can omit `this`, it's much cleaner to write the code like this:

```
with(Calculator()) {
    println(execute(8, Operation.Add(4)))
    println(execute(12, Operation.Multiply(12)))
}
```

> **WARNING** *This is a good time to point out that reading (and writing) text about the `this` reference is a tricky proposition, because the word "this" is going to appear a lot and not be a reference to an object, but instead a simple English word. In this book, you can pay particular attention to the font face. When the word `this` is in code font, it's a reference to an object instance. When this is in a normal font, it's just a part of regular English.*

Your IDE will also give you helpful hints; Figure 11.3 shows what IntelliJ gives you within your `with` lambda expression. This is similar to the indication of what `it` references shown in Figure 11.2 earlier in the chapter.

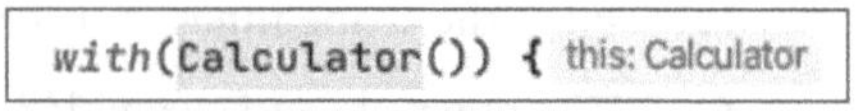

```
with(Calculator()) { this: Calculator
```

FIGURE 11.3: An IDE helping with the this reference in a with expression

The use of `this` instead of `it` is one of the most significant differences between `with` and `let`. It makes a lot of sense, though. You are effectively saying in your code, "With the following object instance, do these things." It would be redundant to then have to say, "With that instance, do this. Then with that instance, do that."

Instead, you tell Kotlin one time, "With the following object instance . . ." and then, within the lambda, say, "Do this" and then "Do that." The `with` sets the context, and everything else in the `with` relates back to that context.

A this Reference Is Always Available

It's important to really get used to having a `this` reference as implied in all code within a `with` block. While it may sound like that's just repeating the previous section, look at the following code:

```
with (Person("David", "Le")) {
    println("My name is ${fullName()}")
}
```

This is legal code and makes a lot of sense. The `with` block takes in a new instance of `Person`, and `this` references that instance. `fullName()` is a method that is being called on the instance through `this`, and since you don't need to explicitly type `this`, the `fullName()` function just appears inside the `println()` seemingly unattached to any instance.

Still, this code just looks . . . weird. That's often the case with a `with` function; there is implicit context that can be confusing. For this reason, be careful using `with`. A little extra commenting might help, because remember, the goal of scope functions is convenience and clarity:

```
with (Person("David", "Le")) {
    // Print out user's name
    println("My name is ${fullName()}")
}
```

> **NOTE** with *gets a helpful assist from good IDEs. Most IDEs will provide syntax highlighting, code completion, and hovering tool tips. All of these make it easier to keep up with the context object and functions available to the block via* this. *Still, you should remember that your code might not be viewed with the same editor you used (or in an editor at all), so commenting and self-documentation remain important.*

All that said, `with` is an important idiomatic programming technique in Kotlin, so don't shy away from using it.

with Returns the Result of the Lambda

Like `let` and most scope functions, `with` returns the result of executing the lambda function attached. That in turn means the result of executing the *last* line of the lambda. Look at the following code:

```
var result = with (Person("David", "Le")) {
    // Print out user's name
    println("My name is ${fullName()}")
}
```

The result of the `with` here is actually the `println()`, which is a `Unit` object instance pointing at the output string.

However, unlike `let`, `with` really isn't that great as a value producer. It's common Kotlin practice to use `with` for calling multiple functions on the context object, and not return or use the result. In other words, `with` is a great shortcut or convenience function for working with an object multiple times and is *not* as great for performing operations that return a single value.

RUN IS A CODE RUNNER AND SCOPE FUNCTION

`run` offers yet another scope function, and it takes on aspects of both `let` and `with` that you've already seen. In many ways, `run` behaves like `let`. It takes a context object and operates upon that context object. However, like `let`, it operates directly upon the context object.

Here's a simple example of using `with` from the last section:

```
with(Calculator()) {
    println(execute(8, Operation.Add(4)))
    println(execute(12, Operation.Multiply(12)))
}
```

Here's the same code using `run`:

```
Calculator().run {
    println(execute(8, Operation.Add(4)))
    println(execute(12, Operation.Multiply(12)))
}
```

Both bits of code output the same results:

```
12
144
```

Choosing a Scope Function Is a Matter of Style and Preference

It may seem odd, but programming is at least partially creative art. Scope functions are a great example of this: you really can use a `with` or a `run`, or even a `let`, all to do the same thing. In fact, here's the same code as just shown except using `let`:

```
Calculator().let {
    println(it.execute(8, Operation.Add(4)))
    println(it.execute(12, Operation.Multiply(12)))
}
```

So, what's the difference? It really does just come down to style. With `let` and `run`, you attach the scope function directly to the context object. `let` uses `it` to reference the context object and `run` uses `this` (and therefore means you can omit the `this` in your code as it's implicit for the compiler). `with` looks more like a `when` control flow, and also uses `this`.

Each looks a bit different than the other. `run` is probably most commonly used when you have an object like `Calculator` that you just want to execute operations against. `run` also returns the value of the lambda, which makes it useful for both initializing an object instance and then storing the result of an operation—a function call—on that object instance.

The following is an example of both initializing an object instance and then storing the result:

```
var product = Calculator().run {
    execute(20435, Operation.Multiply(12042))
}
println(product)
```

Again, to be clear, there's nothing unique about this code, and it could be written with a `let`:

```
// Example of run but using let
Calculator().let {
    it.execute(20435, Operation.Multiply(12042))
}
```

There's nothing wrong here, but it does introduce the `it` reference, which really isn't needed. The context object in `run` uses `this`, which makes the usage with `run` easier to read and more natural.

The same example could also be written with a `with`:

```
// Example of run but using with
with (Calculator()) {
    execute(20435, Operation.Multiply(12042))
}
```

What's the best? None of them. It really comes down to how you want your code to read. However, the Kotlin documentation actually provides some guidance if you still aren't sure. Figure 11.4 shows an excerpt from Kotlin's official page on scope functions regarding the use of run.

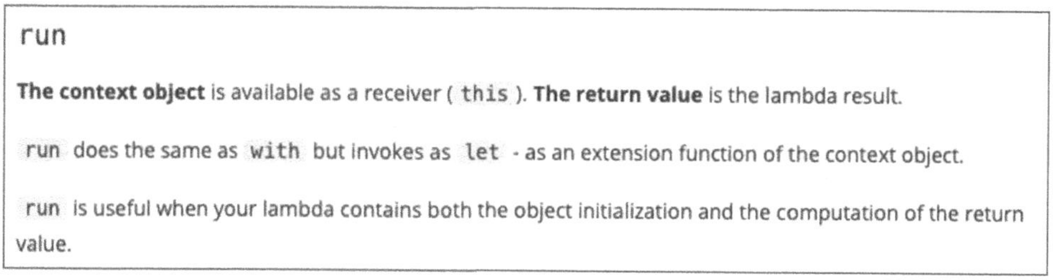

---

run

The context object is available as a receiver (`this`). **The return value** is the lambda result.

`run` does the same as `with` but invokes as `let` - as an extension function of the context object.

`run` is useful when your lambda contains both the object initialization and the computation of the return value.

FIGURE 11.4: Kotlin suggests using run to both initialize an instance and get a result.

There is something very clean about using `run`, because it most closely matches what the code is doing: running some functions and taking the result.

run Doesn't Have to Operate on an Object Instance

One thing that is somewhat unique about `run` is that it doesn't require a context object. The following is an example of that sort of usage:

```
// run without a context object
val hexNumberRegex = run {
    val digits = "0-9"
    val hexDigits = "A-Fa-f"
    val sign = "+-"

    Regex("[$sign]?[$digits$hexDigits]+")
}

for (match in hexNumberRegex.findAll("+1234 -FFFF not-a-number")) {
    println(match.value)
}
```

The result of the `run` lambda gets stored in `hexNumberRegex`, which is used in the following `for` statement.

> **NOTE** *This code example is taken from the Kotlin scope functions documentation available online at* `kotlinlang.org/docs/reference/scope-functions` *.html. This is a bit of an unusual usage of* run *because it doesn't include a context object. There's very little advantage to using* run *like this, as it doesn't add much in terms of readability and saves a negligible amount of memory for variables that only exist within the scope function.*

APPLY HAS A CONTEXT OBJECT BUT NO RETURN VALUE

`apply` is yet another tool in your scope function toolbox and continues the pattern of providing small variations on a theme of convenience and readability. `apply`'s main difference from `let`, `run`, and `with` is that it does not return a value from the lambda expression.

Like `run` and `with`, `apply` uses `this` as the scope object. So `apply` becomes a perfect function for operating on an instance without needing return values back. It's convenient in that you won't need to type the instance name over and over.

The following code uses `apply`:

```
// Examples without using apply
val brian = Person("Brian", "Truesby")
brian.partner = Person("Rose", "Elizabeth", _age = 34)
// For some reason, all we know is that Brian is two years younger than his partner
brian.age = brian.partner?.let { it.age - 2 } ?: 0
println(brian)
println(brian.partner)
```

The thing that might jump out at you is how many times you see the instance variable `brian` appear. It's used a lot!

> **WARNING** *If you're entering and executing the code as you follow along—and you should be!—you'll need to import* org.wiley.kotlin.person.Person *into your project and test program.*

apply Operates upon an Instance

The key here is that after the instance of `Person` is created, the resulting instance, `brian`, is operated upon multiple times. That's where `apply` is useful.

The following is how you could rewrite this code:

```
// Rewriting this code with apply
val brian = Person("Brian", "Truesby").apply {
    partner = Person("Rose", "Elizabeth", _age = 34)
    age = partner?.let { it.age - 2 } ?: 0
    println(this)
    println(partner)
}
```

As has been the case with other scope functions, there's no change in functionality. However, the code is cleaner and clearer. Both `partner` and `age` are referenced through `this`, and clearly are attached to the initial `Person` instance.

apply Returns the Context Object, Not the Lambda Result

Another important thing to note about `apply` is that the return from the function is the instance being operated upon—the context object. So, in this code, the context object is a `Person` instance:

```
// Rewriting this code with apply
val brian = Person("Brian", "Truesby").apply {
    partner = Person("Rose", "Elizabeth", _age = 34)
    // For some reason, all we know is that Brian is two years younger than
his partner
    age = partner?.let { it.age - 2 } ?: 0
    println(this)
    println(partner)
}
```

That means that `brian` at the end holds the `Person` instance created, *with* the updated values from within the `apply` lambda block.

In some ways, you can think of `apply` as an extended `init` block. If you need to create a new instance of an object and configure that object, and all that configuration can be done *within* the instance, then `init` works fine. In this example, though, there's data outside of the `brian` instance involved, such as the age of the `partner`, and some printing that wouldn't make sense to be common for all `Person` instances.

Anytime you have this sort of configuration or instance work that is specific to an instance or doesn't make sense to be a part of an `init` block, `apply` is a great alternative solution.

Also notice this line:

```
println(this)
```

Although not common, you certainly can use the `this` reference to get at the context object directly. That's the need here, as `println()` would not assume a `this` reference as an input argument, so there's a need to explicitly pass `this` into the `println()` function.

?: Is Kotlin's Elvis Operator

There's a slight detour from scope functions worth mentioning. This line of code might look a bit odd to you:

```
age = partner?.let { it.age - 2 } ?: 0
```

First, make sure you understand the scope function pieces of this. The line could be written more explicitly with the `this` reference:

```
this.age = this.partner?.let { it.age - 2 } ?: 0
```

The context object (`brian`, an instance of `Person`) is getting a value for the `age` property. In this (somewhat silly) example, the `brian` instance's `age` needs to be set to 2 less than the partner's age.

This gets immediately tricky, because even though you've assigned another `Person` instance to part-ner in the previous line, `partner` as a property can be `null`. It's declared in `Person` like this:

```
var partner: Person? = null
```

Kotlin knows this, so you need to use a `let` function to ensure that grabbing the `age` from `partner` doesn't fail compilation due to `null` safety.

Then, in the `let` lambda, `it` is used to reference `partner`. The partner's age is retrieved, 2 is subtracted from it, and the result of that expression ends up in `this.age`, which is the age of the `brian` instance.

But there's still a problem. This portion of the expression handles the case when `partner` is not `null`:

```
partner?.let { it.age - 2 }
```

It's easy to think the work is done, because you *know* that `partner` has a value assigned. But Kotlin has to assume, like any good compiler, that other threads might be changing the value of the `partner` property. It is theoretically possible that between your assignment of a `Person` instance to `partner` and this line, `partner` could have been set back to `null`.

For this reason, you've got to provide a non-`null` value to assign to `age` (the receiver of this line of code) in the event that `partner` is `null` and the `let` doesn't execute. You could turn this whole thing into a big `if` statement, but that actually gets tricky, and it's not worth digging into because Kotlin gives you a better option: the *Elvis operator*. The operator is a `?` and then a `:`, like this:

```
?:
```

This allows you to indicate an expression to evaluate and return if the preceding expression is null. Here's that whole line, with the Elvis operator, in context:

```
age = partner?.let { it.age - 2 } ?: 0
```

If `this.partner` evaluates to non-null, the `let` runs and assigns the result of its lambda to `age`. If `partner` is `null`, then the Elvis operator kicks in and returns `0`. You can also have expressions on the right side of the Elvis operator as well.

> **NOTE** *Believe it or not, the Elvis operator is not named for a notable computer scientist or mathematician. Instead, it's named for the so-called king of rock and roll, Elvis Presley. Many think that the question mark (?) followed by a colon (:) looks like a sideways version of Elvis, with two eyes and a huge quiff of hair. Who said programmers aren't trendy?*

ALSO GIVES YOU AN INSTANCE . . . BUT OPERATES ON THE INSTANCE FIRST

The last scope function is `also`. Like the other scope functions, `also` does just what it suggests. It takes an action (in this case, an assignment), and *also* does a few other things.

Here's an example of `also` in action that will reveal some of its key functionality:

```
// Using also to perform actions before assignment
println(Person("David", "Le").also {
    println("New person created: ${it.fullName()}")
    it.partner = Person("Chloe", "Herng Lee").apply {
        println("New partner created: ${fullName()}")
    }
    println("${it.firstName}'s partner is ${it.partner?.fullName()}")
})
```

The output from this code is shown here:

```
New person created: David Le
New partner created: Chloe Herng Lee
David's partner is Chloe Herng Lee
David Le
```

also Is Just Another Scope Function

The first thing to notice is that `also` works in the same basic fashion as you've seen and is particularly similar to `let` and `run`. It is applied to a context object with the dot notation:

```
println(Person("David", "Le").also {
```

Then, as with other scope functions, a lambda expression is provided. The scope function has an `it` reference, which references the context object, the object that `also` is attached to. So here, the context object is a new instance of a `Person`:

```
println(Person("David", "Le").also {
```

Figure 11.5 shows the IDE giving a hint to indicate what `it` references.

```
// Using also to perform actions before assignment
println(Person( firstName: "David",  lastName: "Le").also { it: Person
```

FIGURE 11.5: The also function provides an it reference to the context object.

Within the lambda function, there's some printing and another scope function, mostly to help show the order in which functions execute:

```
println(Person("David", "Le").also {
    println("New person created: ${it.fullName()}")
    it.partner = Person("Chloe", "Herng Lee").apply {
        println("New partner created: ${fullName()}")
    }
    println("${it.firstName}'s partner is ${it.partner?.fullName()}")
})
```

The lambda does some printing using the `it` reference that `also` supplies. In this case, `it` points to the new instance of `Person` with a `firstName` of David and a `lastName` of Le. So the first line is going to print out that name:

```
println("New person created: ${it.fullName()}")
```

The key here is that the initialization of the `Person` instance happens first, and then the `also` block executes. That sounds obvious, but the order of things will become more important shortly.

Then another scope function is used—this time, `apply`:

```
it.partner = Person("Chloe", "Herng Lee").apply {
    println("New partner created: ${fullName()}")
}
```

`apply` uses a `this` reference, so there's no collision between the two scope functions. `it` still references the first `Person`, David Le. In the `apply` block, `this` references the second `Person`, Chloe Herng Lee. So the call to `fullName()` is attached to `this`, and should print out her name.

> **NOTE** *Once you start using scope functions more often, you'll find you use them a lot more often, and it will be common to use one scope function within another. Because most scope functions are close to interchangeable, you may choose one scope function over another because one uses* it *and the other uses* this, *or vice versa. Avoiding reference collision is a perfectly valid reason to choose* apply *over* let, *or* run *over* also.
>
> *This sort of potential for reference collision may also cause you to rename your scope variables, too. But as a rule, you should try to avoid renaming the* this *reference; it's a bad practice that you should avoid if at all possible.*

After the `apply` finishes, there's a bit more printing:

```
println("${it.firstName}'s partner is ${it.partner?.fullName()}")
```

This is within the `also` but outside of the `apply`. `it` references the David Le `Person` instance, which now has a new `partner`, the Chloe Herng Lee instance created and used in the previous `apply`.

This final `println()` is also the last statement executed . . . right?

also Executes before Assignment

Look again at this overall block of code:

```
println(Person("David", "Le").also {
    println("New person created: ${it.fullName()}")
    it.partner = Person("Chloe", "Herng Lee").apply {
        println("New partner created: ${fullName()}")
    }
    println("${it.firstName}'s partner is ${it.partner?.fullName()}")
})
```

At a glance, and based on what you know, it's a fair guess to expect the output to be something like this:

```
David Le
New person created: David Le
New partner created: Chloe Herng Lee
David's partner is Chloe Herng Lee
```

In other words, the code executes from the first line to the last, in order. In that case, the `also` functions as code that runs after the initialization of the instance it's attached to.

Compile this code and run it. You'll get something different:

```
New person created: David Le
New partner created: Chloe Herng Lee
David's partner is Chloe Herng Lee
David Le
```

The output reflects the key difference between `also` and other scope functions: the `also` block executes *prior to* the assignment of the context object to any variable or passing to any function.

Unfortunately, this is a bit counterintuitive, particularly as the scope function is called "also." It is easy to envision `also` as "Initialize (or use) this object instance, and (then) *also* execute this code" but that's not what `also` does. Instead, you should think of `also` as "Execute this block of code, and also assign the context object to the variable (or pass it to the function) once complete with the block."

> **WARNING** *No matter how many different ways you twist or rewrite a description of* also, *it's going to come out a bit awkward. The naming isn't ideal, frankly. Even the official Kotlin documentation stumbles here, saying you can read* also *as "and also do the following with the object." But the "and" in that sentence still implies an order, and that the "doing" (the block of code in the lambda) follows the object assignment.*
>
> *This is a case where you should put less credence in the name of the function as being self-descriptive and just focus on memorizing what it actually does.*

With this in mind, you can make more sense of the ordering of the code. Here's the code again first as a reminder:

```
println(Person("David", "Le").also {
    println("New person created: ${it.fullName()}")
    it.partner = Person("Chloe", "Herng Lee").apply {
        println("New partner created: ${fullName()}")
    }
    println("${it.firstName}'s partner is ${it.partner?.fullName()}")
})
```

The following is exactly what happens, in order:

1. A new `Person` instance is initialized (David Le).

2. That instance is passed into the `also` block:

    a. The `fullName()` function on the `Person` instance (David) is printed.

    b. A new `Person` instance (Chloe Herng Lee) is created.

    c. An `apply` is called on this new `Person` instance (Chloe):

        i. The full name of the `Person` instance (Chloe) is printed.

d. The new `Person` instance (Chloe) is assigned to the `partner` property of the first `Person` instance (David).

e. More printing occurs, displaying the first `Person` instance (David) and then that instance's `partner` property, the newer `Person` instance (Chloe).

3. Finally, the original `Person` instance (David)—along with the updated `partner` property (Chloe)—is passed into the `println()` on the first line of the entire code sample.

All of this happens quickly, and in many cases, the order isn't as noticeable, but it's important to understand that at the compilation and execution levels, there *is* an order to this.

SCOPE FUNCTIONS SUMMARY

Now that you've seen all these scope functions, you've realized the most important thing about them: they're very similar and almost interchangeable. Using one over the other has more to do with style and preference than a particular use case.

There are a few key differences that will help you separate each scope function from the others:

➤ The name of the reference variable, `it` or `this` (even though you can override the name for all scope functions).

➤ What is returned: the context object or the result of the lambda function.

➤ Whether the scope function is an extension function, and therefore attached to the context object with dot notation.

Once you have each scope function aligned with these three items, you really have a complete picture, and can pick and choose at will. Table 11.1 summarizes the differences.

TABLE 11.1: Differences between scope functions

SCOPE FUNCTION	CONTEXT OBJECT REFERENCE	RETURN VALUE FROM FUNCTION	USAGE
let	it	Lambda result	Extension function (dot notation on context object)
with	this	Lambda result	Context object passed in
run	this (or nothing)	Lambda result	Extension function or without a context object
apply	this	Context object	Extension function
also	it	Context object	Extension function

Listing 11.9 helps show the differences. It's a complete listing of the various scope function samples used throughout the chapter.

LISTING 11.9: The complete set of samples for scope functions

```kotlin
import org.wiley.kotlin.math.Calculator
import org.wiley.kotlin.math.Operation
import org.wiley.kotlin.person.Person
import org.wiley.kotlin.user.SimpleUser
import org.wiley.kotlin.user.User

fun main() {
    var result = Calculator().let { calc ->
        var intFunction = { 5 - 2 }
        println(calc.add(intFunction(), 3, 5, 3))
        println(calc.execute(5, Operation.Add(4)))
        println(calc.execute(2, Operation.Add(4)))

        calc.execute(5, Operation.Multiply(5)).let { result ->
            println("Inner result: ${calc.execute(result, Operation.Add(12))}")
        }
    }

    println(result)

    var calc = Calculator()
    theFun(calc)

    var bigResult = Calculator().let { calc ->
        calc.execute(8, Operation.Add(4)).let {
            calc.execute(12, Operation.Multiply(it))
        }.let {
            calc.execute(it, Operation.Divide(18))
        }.let {
            calc.execute(16, Operation.Subtract(it))
        }.let {
            calc.execute(it, Operation.Raise(2))
        }
    }

    println("Big result is ${bigResult}")

    var nestedResult = Calculator().let { calc ->
        calc.execute(8, Operation.Add(4)).let {
            calc.execute(12, Operation.Multiply(it)).let {
                calc.execute(it, Operation.Divide(18)).let {
                    calc.execute(16, Operation.Subtract(it)).let {
                        calc.execute(it, Operation.Raise(2))
                    }
                }
            }
        }
    }
```

continues

LISTING 11.9 *(continued)*

```kotlin
    println("Nested result is ${bigResult}")

    var calculator = Calculator()
    var one = calc.execute(8, Operation.Add(4))
    var two = calc.execute(12, Operation.Multiply(one))
    var three = calc.execute(two, Operation.Divide(18))
    var four = calc.execute(16, Operation.Subtract(three))
    var five = calc.execute(four, Operation.Raise(2))

    println(five)

    with(Calculator()) {
        println(execute(8, Operation.Add(4)))
        println(execute(12, Operation.Multiply(12)))
    }

    var withResult = with (Person("David", "Le")) {
        println("My name is ${fullName()}")
    }
    println(withResult)

    // run example
    Calculator().run {
        println(execute(8, Operation.Add(4)))
        println(execute(12, Operation.Multiply(12)))
    }

    // run example but using let
    Calculator().let {
        println(it.execute(8, Operation.Add(4)))
        println(it.execute(12, Operation.Multiply(12)))
    }

    // "Typical" example of run
    var product = Calculator().run {
        execute(20435, Operation.Multiply(12042))
    }
    println(product)

    // Example of run but using let
    Calculator().let {
        it.execute(20435, Operation.Multiply(12042))
    }

    // Example of run but using with
    with (Calculator()) {
        execute(20435, Operation.Multiply(12042))
    }

    // run without a context object
    val hexNumberRegex = run {
        val digits = "0-9"
        val hexDigits = "A-Fa-f"
```

```
            val sign = "+-"

            Regex("[$sign]?[$digits$hexDigits]+")
        }

        for (match in hexNumberRegex.findAll("+1234 -FFFF not-a-number")) {
            println(match.value)
        }

        // Examples without using apply
        /*
        val brian = Person("Brian", "Truesby")
        brian.partner = Person("Rose", "Elizabeth", _age = 34)
        // For some reason, all we know is that Brian is two years younger than
    his partner
        brian.age = brian.partner?.let { it.age - 2 } ?: 0
        println(brian)
        println(brian.partner)
         */

        // Rewriting this code with apply
        val brian = Person("Brian", "Truesby").apply {
            partner = Person("Rose", "Elizabeth", _age = 34)
            // For some reason, all we know is that Brian is two years younger than
    his partner
            age = partner?.let { it.age - 2 } ?: 0
            println(this)
            println(partner)
        }

        // Using also to perform actions before assignment
        println(Person("David", "Le").also {
            println("New person created: ${it.fullName()}")
            it.partner = Person("Chloe", "Herng Lee").apply {
                println("New partner created: ${fullName()}")
            }
            println("${it.firstName}'s partner is ${it.partner?.fullName()}")
        })
    }

    fun theFun(calc: Calculator) {
        var intFunction = { 5 - 2 }
        println(calc.add(intFunction(), 3, 5, 3))
        println(calc.execute(5, Operation.Add(4)))
        println(calc.execute(2, Operation.Add(4)))
    }
```

It's a *little* bit of overstatement to say that it's all about style in choosing a scope function, but it drives the point home. also in particular operates a bit differently in that it executes before assignment, and you can argue the case that some returning the context object and others returning the lambda result is a big difference.

The important point is that there's no hard and fast "right time to use such-and-such scope function." The sooner you accept that, the freer you'll be in coding with them.

12

Inheritance, One More Time, with Feeling

Abstract classes

➤ Interfaces for defining behavior

➤ Implementing interfaces

➤ Implementation versus extension

➤ Delegation patterns

➤ Delegation versus implementation

ABSTRACT CLASSES REQUIRE A LATER IMPLEMENTATION

You've already spent a couple of chapters learning about classes and inheritance, and you've built a *lot* of open classes and extended those classes. Now it's time to take one more pass through classes, inheritance, and some concepts you've only briefly seen as well as a few that you've not seen yet at all.

First is the abstract class. You learned in Chapter 8 that data classes cannot be abstract, even though you had not learned much about abstract classes at that point. You also briefly saw an abstract method in Chapter 9:

```
abstract fun isSuperUser(): Boolean
```

abstract in Kotlin is in many ways a deferral term. It says, "This defines functionality that will have to be implemented later." In the case of a method, it requires a class that extends the class to override that method. In the case of a class, it means that the class cannot be directly instantiated.

Abstract Classes Cannot Be Instantiated

Listing 12.1 revisits the `User` class shown in several previous chapters. This is a data class, and has a single function explicitly defined. As it's a data class, it will also handle property access to `email`, `firstName`, and `lastName`.

LISTING 12.1: The User data class is a class for extension

```
package org.wiley.kotlin.user

data class User(var email: String, var firstName: String, var lastName: String) {

    override fun toString(): String {
        return "$firstName $lastName with an email of $email"
    }
}
```

> **NOTE** *You might remember that data classes cannot be abstract. As you'll see, in many ways, a data class is the opposite of an abstract class. Data classes are all about implementing behavior, and abstract classes are primarily concerned with defining behavior, but not implementing it.*

If you think about `User`, it's a great candidate for being a base class. You can imagine a number of extensions: an `Administrator`, a `SuperUser`, a `GuestUser`, and so on. Each of these extensions could extend `User` and add type-specific behavior.

The same is true for the `Band` class defined and used in previous chapters, and the `Person` class as well. These classes are perfect candidates for extension and then overriding behavior or adding type-specific behavior.

What's particularly important to understand, though, is that the base classes—`User`, `Band`, and `Person`—can stand on their own. There is absolutely the concept of a general `User`, a `Band`, and a `Person`. It makes sense in many contexts to instantiate `User`, `Band`, and `Person`.

In contrast, look at Listing 12.2, a new class called `Car`.

LISTING 12.2: A basic Car class

```
package org.wiley.kotlin.car

class Car {

    fun drive(direction: String, speed: Int) : Int {
        // drive in the given direction at the given speed
        // return the distance driven — fix later
        return 0
    }
}
```

> **WARNING** *In Chapter 9, you created an* Auto *class in the* org.wiley.kotlin. auto *package. This chapter uses a new package,* org.wiley.kotlin.car, *and a new class name,* Car, *to distinguish from that earlier class.*

At first glance, this seems like yet another variation on a theme: a base class that can be extended into Porsche and Audi and Ford and Honda subclasses, each with their own unique behavior, and likely some amount of shared behavior as well.

However, there's an important difference between Car as opposed to Band or User. There really is no such thing as a "generic" Car. Every Car is a particular make (and you can extend this to model if you want). However, a Band *is* a standalone concept. An instance of Band makes sense, because it is self-defining. A RockBand instance might be more specific to a type of music, but a plain old Band instance also works.

That's not the same as Car. A Car has to have certain things. When this is the case, you really don't want an instance of a Car. It's incomplete. And you can use Kotlin to disallow direct creation of Car instances by creating an abstract class, rather than a non-abstract class—regardless of whether that class is open or not.

> **NOTE** *If you've just thought of 10 or 20 objections to this line of reasoning, that's great! They may be valid—just keep reading, as this is a complex and somewhat nuanced topic that will take more than a few paragraphs to explain fully.*

Listing 12.3 converts Car to an abstract version.

LISTING 12.3: An abstract version of the Car class

```
package org.wiley.kotlin.car

abstract class Car {

    fun drive(direction: String, speed: Int) : Int {
        // drive in the given direction at the given speed
        // return the distance driven - fix later
        return 0
    }
}
```

Believe it or not, just adding abstract before the first line of the class definition is all that's required.

You can see the effect of this by creating a new sample program, like the one shown in Listing 12.4.

LISTING 12.4: Testing out an abstract class

```
import org.wiley.kotlin.car.Car

fun main() {
    var car = Car()
}
```

You'll immediately get a compiler error:

```
Kotlin: Cannot create an instance of an abstract class
```

This is pretty self-explanatory. You can't instantiate an abstract class. More importantly, though, it gives you a sense of what abstract classes are really for: defining must-have behavior without defining the behavior itself.

Abstract Classes Define a Contract with Subclasses

Car defines a sort of contract with anything that subclasses it. However, right now, that contract is empty. You can't instantiate Car, but if you subclass it, you don't have to fulfill any requirements. That makes Car a rather useless abstract class.

A better approach is to define behavior in the abstract class and then defer implementation to subclasses. Listing 12.5 shows this and makes the drive() function abstract as well.

LISTING 12.5: Making the single function in Car abstract

```
package org.wiley.kotlin.car

abstract class Car {

    // Subclasses need to define how they drive
    abstract fun drive(direction: String, speed: Int) : Int
}
```

In this case, even though the Car class is abstract overall, it's an effect. The class being abstract is the effect of having at least one abstract function. This is an important thing to realize: abstract classes must be extended specifically so that the abstract functions in those abstract classes can be overridden and given behavior.

This is the contract that you'll often hear, and it's a contract between the defining class—the abstract class, in this case Car—and any and all classes that subclass Car. The abstract class in this case says that this method must be defined with this exact signature:

```
// Subclasses need to define how they drive
abstract fun drive(direction: String, speed: Int) : Int
```

Now classes that extend Car know what is required of them. Listing 12.6 shows a simple extension (with the emphasis on simple) of the Car class that fulfills this contract.

LISTING 12.6: Fulfilling the contract laid out by Car

```
package org.wiley.kotlin.car

class Honda : Car() {

    override fun drive(direction: String, speed: Int) : Int {
        // Demo only. Super simple: just return speed / minutes in an hour
        return speed / 60
    }
}
```

Listing 12.7 uses Honda in an equally simple example.

LISTING 12.7: Instantiating Honda and using the drive() function

```
import org.wiley.kotlin.car.Honda

fun main() {
    var car = Honda()
    println("The car drove ${car.drive("W", 60)} miles in the last minute.")
}
```

The output is what you'd expect:

```
The car drove 1 miles in the last minute.
```

> **WARNING** *Fair warning: the implementation of functions in this chapter will likely neither impress nor amaze you. That's OK; the goal isn't to write particularly tricky code, but to really focus on the various approaches to inheritance that Kotlin provides, first with abstract classes and functions and soon with interfaces and implementations of those interfaces. Don't get too overly worried about the function code, but instead pay attention to how these classes and constructs interact, and why inheritance works the way it does in Kotlin.*

Abstract classes usually do more than just define a single abstract function, though. It's common to see a number of functions as part of the contract laid out by the abstract class. Abstract classes can also define properties that need to be handled by subclasses. Listing 12.8 shows a slightly evolved version of Car that adds an abstract property and some additional abstract functions.

LISTING 12.8: Adding more details to the contract that Car defines

```
package org.wiley.kotlin.car

abstract class Car {
```

(continues)

LISTING 12.8 *(continued)*

```
    abstract var maxSpeed: Int

    // Start the car up
    abstract fun start()

    // Stop the car
    abstract fun stop()

    // Subclasses need to define how they drive
    abstract fun drive(direction: String, speed: Int) : Int
}
```

> **WARNING** *If you're following along and compiling as you go, this change to* Car *will break the* Honda *class. That class no longer fulfills the contract that* Car *lays out because it doesn't handle the new* maxSpeed *property, or the* start() *or* stop() *functions.*

Abstract Classes Can Define Concrete Properties and Functions

So far, everything in Car is abstract: its properties and functions. But there's no requirement for that. In fact, this will turn out to be a key difference between an abstract class and an interface, which you'll learn about later in this chapter. An abstract class can have fully implemented (sometimes called "concrete") functions and properties mixed in with abstract ones.

Change the first line of Car to look like this:

```
abstract class Car(val model: String, val color: String) {
```

Now, every Car subclass has a model and color property available. Those can also be used in the abstract class, even though they won't be defined until compile time, and in a concrete subclass. Listing 12.9 adds a simple toString() function—one that is concrete—to Car. This method is now available to all subclasses of Car.

LISTING 12.9: Setting up some concrete functions and properties in Car

```
package org.wiley.kotlin.car

abstract class Car(val model: String, val color: String) {

    abstract var maxSpeed: Int

    // Start the car up
    abstract fun start()

    // Stop the car
    abstract fun stop()
```

```
    // Subclasses need to define how they drive
    abstract fun drive(direction: String, speed: Int) : Int

    override fun toString() : String {
        return "${this::class.simpleName} ${model} in ${color}"
    }
}
```

> **NOTE** toString() *here uses the* this *reference, which you know about, and the* ::class *notation to get access to the current class (at runtime). Then, the* simpleName *property accesses that class's name. This is a means of getting at whatever class is actually instantiated, be it* Honda, Porsche, *or something else— at runtime.*

Listing 12.10 is an updated version of Honda that implements all the functions and the property required by Car, and Listing 12.11 is an update of the sample code to run to test this all out.

LISTING 12.10: Updating Honda to fulfill the contract set out by Car

```
package org.wiley.kotlin.car

class Honda(model: String, color: String) : Car(model, color) {

    override var maxSpeed : Int = 128

    override fun start() {
        println("Starting up the Honda ${model}!")
    }

    override fun stop() {
        println("Stopping the Honda ${model}!")
    }

    override fun drive(direction: String, speed: Int) : Int {
        println("The Honda ${model} is driving!")
        return speed / 60
    }
}
```

LISTING 12.11: A simple program to show off Honda

```
import org.wiley.kotlin.car.Honda

fun main() {
    var car = Honda("Accord", "blue")
    car.start()
    car.drive("W", 60)
    car.stop()
}
```

Subclasses Fulfill the Contract Written by an Abstract Class

Now you see that an abstract class is of no real value unless and until it is subclassed. But it really goes further: most abstract classes are intended to be subclassed by *multiple* subclasses. You wouldn't create `Car` unless you intended more than just a `Honda` subclass.

Subclasses Should Vary Behavior

When you start to see multiple subclasses, though, there are some things to be careful about. First, you really want subclasses to have different behaviors. To see this, look first at a counterexample. Listing 12.12 creates another subclass of `Car` called `Porsche`.

LISTING 12.12: Building another subclass of Car

```
package org.wiley.kotlin.car

class Porsche(model: String, color: String) : Car(model, color) {

    override var maxSpeed : Int = 212

    override fun start() {
        println("Starting up the Porsche ${model}!")
    }

    override fun stop() {
        println("Stopping the Porsche ${model}!")
    }

    override fun drive(direction: String, speed: Int) : Int {
        println("The Porsche ${model} is driving!")
        return speed / 60
    }
}
```

This is actually a poor example of subclassing—not just the `Porsche` class, but in combination with the `Honda` class. Both `Honda` and `Porsche` are doing the exact same things in the `start()`, `stop()`, and `drive()` functions.

You could actually refactor these and make these functions concrete in `Car`:

```
// Start the car up
fun start() {
    println("Starting up the ${this::class.simpleName} ${model}!")
}

// Stop the car
fun stop() {
    println("Stopping the ${this::class.simpleName} ${model}!")
}
```

At this point, the base class—abstract or not—takes care of printing a generic message associated with starting and stopping each specific instance type.

Imagine for the sake of example that the details of how these various cars start is different. That's not actually far-fetched, and the further you go into modeling the behavior of how the vehicles actually start up, the truer this becomes. So you might have this for how a Honda starts:

```
override fun start() {
    println("Inserting the key, depressing the brake, pressing the ignition,
starting.")
}
```

And then you might have this for the Porsche:

```
override fun start() {
    println("Remote starting, depressing the brake, shifting into drive")
}
```

You could imagine replacing those println() statements with other function calls to various vehicle functions.

In general, if your subclasses all have the same behavior in them, then you should revisit how you are using inheritance.

The Contract Allows for Uniform Treatment of Subclasses

As you develop your inheritance tree, you should see more and more subclasses, but a greater degree of variety within how those subclasses override the base abstract class's behavior.

The key is what is common: all subclasses originate in a single base class. That means that you can treat the subclasses *as* the base class and operate upon the *contract* rather than the specific subclass. Now that's a lot of formal language, so it's worth walking through how that looks.

First, you need a few more Car subclasses. Listings 12.13 and 12.14 add an Infiniti and a BMW to the mix.

LISTING 12.13: An Infiniti subclass of Car

```
package org.wiley.kotlin.car

class Infiniti(model: String, color: String) : Car(model, color) {

    override var maxSpeed : Int = 167

    override fun start() {
        println("Starting the Infiniti ${model}: Inserting the key, starting
the engine")
    }

    override fun stop() {
        println("Stopping the Infiniti ${model}!")
    }

    override fun drive(direction: String, speed: Int) : Int {
        println("The Infiniti ${model} is driving!")
        return speed / 60
    }
}
```

LISTING 12.14: A BMW subclass of Car

```kotlin
package org.wiley.kotlin.car

class BMW(model: String, color: String) : Car(model, color) {

    override var maxSpeed : Int = 182

    override fun start() {
        println("Starting the BMW ${model}: Depressing the brake, pushing the
starter button")
    }

    override fun stop() {
        println("Stopping the BMW ${model}!")
    }

    override fun drive(direction: String, speed: Int) : Int {
        println("The BMW ${model} is driving!")
        return speed / 60
    }
}
```

> **NOTE** *As mentioned earlier, these are pretty trivial examples of* Car *subclasses without as much variation as you'd want in a more realistic application. The focus here is on minor printing differences so it's easy to see which subclass is being invoked in the test class, shown in Listing 12.15.*

Now you can build a program that loads up a bunch of Car subclasses, and then iterates over them, treating each generically . . . even though each subclass has unique behavior. That's exactly what Listing 12.15 does.

LISTING 12.15: Iterating over Cars in a list

```kotlin
import kotlin.random.Random
import org.wiley.kotlin.car.BMW
import org.wiley.kotlin.car.Car
import org.wiley.kotlin.car.Honda
import org.wiley.kotlin.car.Infiniti
import org.wiley.kotlin.car.Porsche

fun main() {
    var car = Honda("Accord", "blue")
    car.start()
    car.drive("W", 60)
    car.stop()
```

```
        println("\nLoading fleet of cars...")
        var cars : MutableList<Car> = loadCars(10)
        for (car in cars) {
            car.start()
            car.drive("E", Random.nextInt(0, 200))
            car.stop()
        }
    }

    fun loadCars(numCars: Int) : MutableList<Car> {
        var list = mutableListOf<Car>()

        for (i in 0..numCars) {
            when (Random.nextInt(0, 3)) {
                0 -> list.add(Honda("CRV", "black"))
                1 -> list.add(Porsche("Boxster", "yellow"))
                2 -> list.add(BMW("435i", "blue"))
                3 -> list.add(Infiniti("QX60", "silver"))
            }
        }
        return list
    }
```

The key here isn't the behavior or the output. It's that from the program's point of view, every entry in the cars list is just a Car. This is where that contract is so important: each Car subclass has a drive() and a start() and a stop(). It really doesn't matter how those functions are implemented. Your code can just trust that they *are* implemented, because that's what the contract laid out by Car establishes.

INTERFACES DEFINE BEHAVIOR BUT HAVE NO BODY

You've seen that abstract classes have a couple of specific properties:

➤   They define a contract for later behavior by requiring that certain functions and properties be implemented

➤   They can add concrete properties and functions that are inherited by subclasses

Interfaces are similar, but can *only* define behavior and properties. They cannot define anything concrete in terms of using any properties for state or being accessed by the rest of the class.

> **NOTE** *That last sentence is a bit odd. What does it mean to use a property for state, or to access that property in the rest of the class? It's okay if that was hard to follow, because it's something quite unusual about interfaces. Keep reading and it will become clearer, but don't worry if you don't quite understand what that means just yet.*

Listing 12.16 is a `Vehicle` interface that looks quite similar to your existing `Car` abstract class.

LISTING 12.16: Building an interface for vehicles

```
package org.wiley.kotlin.vehicle

interface Vehicle {
    val maxSpeed: Int

    fun start()
    fun stop()
    fun drive(direction: String, speed: Int) : Int
}
```

Obviously, there are some differences between this and the `Car` abstract class in Listing 12.9:

➤ An interface uses the `interface` keyword instead of `class`, and an abstract class uses the `class` keyword after `abstract`.

➤ An interface's functions and properties don't have to be preceded by the `abstract` keyword, while an abstract class's functions and properties do.

➤ An interface cannot have constructors or properties taken in as part of initialization, while abstract classes can.

A class can implement an interface as shown in Listing 12.17.

LISTING 12.17: Implementing the Vehicle interface

```
package org.wiley.kotlin.car

import org.wiley.kotlin.vehicle.Vehicle

class Volkswagen : Vehicle {
    override val maxSpeed = 190

    override fun start() {
        println("Starting the VW: Turning the key, shifting into drive")
    }

    override fun stop() {
        println("Stopping the VW")
    }

    override fun drive(direction: String, speed: Int) : Int {
        println("The VW is driving!")
        return speed / 60
    }
}
```

NOTE *The terminology is a bit different here. A class implements an interface but subclasses or extends a class. As you'll see shortly, both of these can go on at the same time, so getting the terminology correct is important.*

It's pretty obvious that Volkswagen, while implementing Vehicle instead of subclassing Car, looks very similar to the other Car classes. In fact, Car and Vehicle are really similar.

Interfaces and Abstract Classes Are Similar

It's no accident that these look alike. In many cases, it is entirely a personal preference as to whether you use an interface or an abstract class. You can also use a bit of both! For example, you could redefine Car to look like Listing 12.18.

LISTING 12.18: An abstract class implementing an interface

```kotlin
package org.wiley.kotlin.car

import org.wiley.kotlin.vehicle.Vehicle

abstract class Car(val model: String, val color: String) : Vehicle {

    // abstract var maxSpeed: Int

    // Start the car up
    // abstract fun start()

    // Stop the car
    // abstract fun stop()

    // Subclasses need to define how they drive
    // abstract fun drive(direction: String, speed: Int) : Int

    override fun toString() : String {
        return "${this::class.simpleName} ${model} in ${color}"
    }
}
```

Most of the code is commented so that you can see that the big changes in Car are removals. It now implements Vehicle, adds details related to a model and color, and defines a toString() function. The function definitions are no longer needed because the maxSpeed property as well as start(), stop(), and drive() are all defined now in Vehicle, and don't need to be redefined in the Car abstract class.

This is a pretty common pattern: define a very generic interface, and then subdivide implementations through using an abstract class. For instance, consider an inheritance hierarchy as visualized in Figure 12.1.

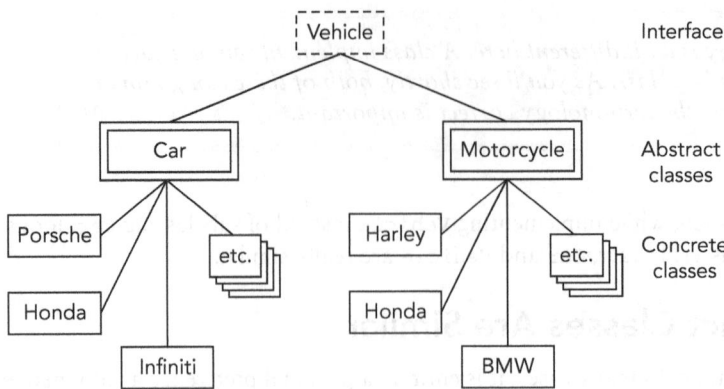

FIGURE 12.1: An inheritance hierarchy using interfaces, abstract classes, and concrete classes

At the top of the hierarchy is a very generic interface, `Vehicle`. It defines required behavior for a broad class—vehicles. Then two abstract classes, `Car` and `Motorcycle`, both implement the interface and then add specificity and refine what needs to be implemented. Then a number of subclasses extend the abstract class (which by extension means that they are implementations of the `Vehicle` interface) and fill in all the required behavior.

This has all the makings of a sound inheritance tree. It begins at the top of the tree with general functionality. At each level, moving down the tree, classes become more specific and the variation becomes more defined. However, all of these classes are ultimately `Vehicle` implementations.

You could have a list of `Vehicle` instances and each item in that list could be a `Car` subclass or a `Motorcycle` subclass, and you could call any function defined in `Vehicle` on each item in that list, knowing that the contract that `Vehicle` creates is being fulfilled.

You could also create a list of just subclasses of `Motorcycle`, and call functions that are specific to `Motorcycle`. You might have `Motorcycle` add a `kickstand` property that a `Car` doesn't have, and any subclass of `Motorcycle` is now going to support that property.

Interfaces Cannot Maintain State

At the beginning of this section on interfaces, you read that an interface can define properties, but they can't define state or be used in the rest of the class. To understand this, take a close look at how `Vehicle` defines the `maxSpeed` property:

```
interface Vehicle {
    val maxSpeed: Int

    // other code
}
```

`maxSpeed` is effectively abstract. If you look back at how this was defined in `Car`, it was marked as abstract there:

```
abstract class Car(val model: String, val color: String) : Vehicle {

    abstract var maxSpeed: Int
```

```
    // other code
}
```

In both cases, the actual handling of `maxSpeed` is deferred to concrete classes. Because of that, `maxSpeed` *can* be used in functions, like `toString()` in `Car`:

```
override fun toString() : String {
    return "${this::class.simpleName} ${model} in ${color} with a max speed of
${maxSpeed}"
}
```

This is legal because Kotlin's compiler understands that by the time this function is called, code in a concrete class will have ensured that `maxSpeed` is populated.

A Class's State Is the Values of Its Properties

A class like BMW has state stored in its property values. Its model state, its color state, and so on, are all stored in `model` and `color` and `maxSpeed` properties:

```
class BMW(model: String, color: String) : Car(model, color) {

    override var maxSpeed : Int = 182
```

A specific instance of BMW will have different state than another specific instance. One might be a blue 435i and another a white M3.

An interface, however, cannot store state. That must be handled by concrete instances. But an interface can do something a bit odd: it can define getters for properties that are *not* deferred until later.

An Interface Can Have Fixed Values

In an interface, you can define a property that is then immediately provided with a custom getter implementation. You can see this in Listing 12.19.

LISTING 12.19: Adding a property with a getter to Vehicle

```
package org.wiley.kotlin.vehicle

interface Vehicle {
    val maxSpeed: Int

    val purpose: String
      get() = "This interface models vehicles of all kinds"

    fun start()
    fun stop()
    fun drive(direction: String, speed: Int) : Int
}
```

This isn't state. It's essentially a fixed value that is now available on all implementations of `Vehicle`. Every instance of `Car` and `Motorcycle` will have a `purpose` property, and the value for all of them when called is the string defined in `Vehicle`:

```
println(car.purpose)
```

You won't find these properties with fixed values defined that often, because you typically want properties to be stateful—to change with each instance. This is sort of like a constant for all interface implementations, which turns out to not be useful most of the time.

> **NOTE** *There is one pretty good use case for fixed-value properties in an interface. You'll see this use case in the section on interfaces that implement other interfaces, coming up shortly.*

Interfaces Can Define Function Bodies

One thing that interfaces in Kotlin can do that may surprise you is define a function *and* that function's behavior. That may seem a little odd; an interface has no properties that define state, so what would a function do? Well, you've actually already seen a good example of this. Remember this version of `toString()` in the abstract `Car` class:

```
override fun toString() : String {
    return "${this::class.simpleName} ${model} in ${color}"
}
```

You could come up with something similar without the `model` property for the `Vehicle` interface:

```
fun stats(): String {
    return "This vehicle is a ${this::class.simpleName} and has a maximum speed of
${maxSpeed}"
}
```

Now this may look a bit odd. At first glance, isn't this breaking the rule about an interface maintaining state? No, because it's not actually the interface maintaining state; it's the concrete implementation somewhere down the inheritance chain. That actual instance will have its own `maxSpeed` that must be handled—because the `maxSpeed` property is defined in the `Vehicle` interface but implemented elsewhere.

In fact, go back to your sample code (last shown in Listing 12.15), and make this addition to the loop that goes through the list of `Car` instances:

```
var cars : List<Car> = loadCars(10)
for (car in cars) {
    println(car.stats())
    car.start()
    car.drive("E", Random.nextInt(0, 200))
    car.stop()
}
```

> **WARNING** *If this code doesn't compile for you, make sure you've taken all the steps detailed in this section. You'll need to have created the Vehicle* interface, *updated* Car *to implement* Vehicle, *and then commented out the functions in* Car *that are now inherited from* Vehicle. *Once all that's done, then any instance of a* Car *subclass will have access to the new* stats() *function defined up in* Vehicle.

This is actually a pretty important point to make: an interface (as well as an abstract class) can quickly add functionality to *all* inheriting classes in this manner. Even with multiple subclasses of Car, Car itself, and even Motorcycle and Motorcycle subclasses, adding a function to Vehicle provided new behavior for *all* of those implementing classes.

Interfaces Allow Multiple Forms of Implementation

You know that a class—even an abstract class—can implement an interface. But an interface can also implement an interface. For example, you could decide that Car is better as an interface than it is as an abstract class, and rework it, as shown in Listing 12.20.

LISTING 12.20: Car as an Interface (with errors)

```
package org.wiley.kotlin.car

import org.wiley.kotlin.vehicle.Vehicle

interface Car(val model: String, val color: String) : Vehicle {

    override fun toString() : String {
        return "${this::class.simpleName} ${model} in ${color}"
    }
}
```

This is going to generate a number of errors, though. First, an interface can't have a constructor, so you'd lose the model and color properties. You could declare them as properties, just as maxSpeed was declared on Vehicle, and then implement them in subclasses, but you lose a lot of nice common behavior—basically, the reason Car was an abstract class in the first place.

Try out Listing 12.21, instead. This defines a new interface called Manufacturer. This could then be extended to the class in Listing 12.22, a HondaManufacturer.

LISTING 12.21: A new interface for vehicle manufacturers

```
package org.wiley.kotlin.vehicle

interface Manufacturer {
    val name: String
}
```

LISTING 12.22: Extending the Manufacturer implementation

```
package org.wiley.kotlin.vehicle

interface HondaManufacturer : Manufacturer {
    override val name: String
        get() = "Honda"
}
```

There is a lot of interesting detail in these two very short interfaces. First, you see that an interface can extend another interface. Second, you're seeing the use case where having a property defined with a fixed `get()` function makes sense: in the case where you have an interface extend another interface.

> **NOTE** *Here is another important naming and lexical convention. A class that inherits from an interface is said to implement the interface. An interface that inherits from another interface is said to extend the interface. That's because the inheriting interface actually doesn't implement behavior; it just adds or extends existing definitions, as is the case with* `HondaManufacturer` *extending* `Manufacturer`*.*
>
> *It's also almost always safe to say that one class or interface inherits from another class or interface. If you're not sure, "inherits from" is usually a safe bet.*

In the top-level interface, `Manufacturer`, a property called name is declared. But `HondaManufacturer` doesn't want to leave that property to be defined later; it wants to set it for all implementing classes of `HondaManufacturer`. So this is where setting the value through a fixed `get()` method *does* make sense. Now classes that implement `HondaManufacturer` don't need to deal with defining name and they all get "Honda" as a name property automatically.

A Class Can Implement Multiple Interfaces

Now you have an interesting situation. The `org.wiley.kotlin.car.Honda` class—which, to be clear, is distinct from a potential `org.wiley.kotlin.motorycle.Honda` class—implements `Car`. But it could, and should, also implement `HondaManufacturer`. That is perfectly legal, as Listing 12.23 shows.

LISTING 12.23: Honda can extend a class and implement interfaces

```
package org.wiley.kotlin.car

import org.wiley.kotlin.vehicle.HondaManufacturer

class Honda(model: String, color: String) : Car(model, color), HondaManufacturer {

    override var maxSpeed : Int = 128

    override fun start() {
        println("Inserting the key, depressing the brake, pressing the ignition,
starting.")
    }
```

```
        override fun stop() {
            println("Stopping the Honda ${model}!")
        }

        override fun drive(direction: String, speed: Int) : Int {
            println("The Honda ${model} is driving!")
            return speed / 60
        }
    }
}
```

From what's visible here, Honda now extends Car and implements HondaManufacturer. But it's also technically implementing Vehicle via Car as well. There's a lot going on in this class!

You can now create an instance of Honda and print out its name property, which it gets from Honda-Manufacturer:

```
var honda = Honda("Odyssey", "grey")
println(honda.name)
```

Interface Property Names Can Get Confusing

You may see a problem here. When looking at just the Manufacturer interface, the name property made plenty of sense. But now, looking at the Honda class, it's not at all clear what name references. You have two simple options to address this:

➤ Prefix an interface's potentially generic property names (like name) with the interface, so use manufacturerName instead of name, for example.

➤ Avoid generic property names, so use manufacturer or make instead of name.

Either is okay, and this becomes a matter of style and preference. For this example, you can simply change the name of the name property in Manufacturer and HondaManufacturer to make, which gives better context for the property's value and purpose.

Interfaces Can Decorate a Class

Something else you might have noticed is that even though Honda (and other subclasses of Car or even the fictional Motorcycle abstract class) can now implement HondaManufacturer, no additional behavior had to be implemented.

This is typically known as decoration, or the decorator pattern. You can read more about this design pattern in a number of places online, but a good Java-specific (and therefore closely related to Kotlin) explanation is here: dzone.com/articles/gang-four-%E2%80%93-decorate-decorator. The idea is that a class or interface *decorates* an existing class with existing data or behavior without having to refactor or impact the existing object.

In this case, HondaManufacturer decorates Honda by giving it a new property, and the Honda class only has to add in another inherited interface. You could also add functionality in this same manner; you might add a function that provides information about the manufacturer. So Manufacturer might define a new function, as shown in Listing 12.24, and then HondaManufacturer could further define behavior as shown in Listing 12.25.

LISTING 12.24: Defining a new function on Manufacturer implementations

```
package org.wiley.kotlin.vehicle

interface Manufacturer {
    val make: String

    fun manufacturerInformation() : String
}
```

LISTING 12.25: Adding behavior available to all implementations of the interface

```
package org.wiley.kotlin.vehicle

interface HondaManufacturer : Manufacturer {
    override val make: String
        get() = "Honda"

    override fun manufacturerInformation(): String {
        return "Honda was the eighth largest automobile manufacturer in the world
in 2015." +
                "Learn more at https://www.honda.com"
    }
}
```

DELEGATION OFFERS ANOTHER OPTION FOR EXTENDING BEHAVIOR

At this point, you've gotten the bulk of "everyday inheritance" in Kotlin down. Nine times out of ten, you'll use interfaces, and either implement them with abstract or concrete classes or extend them with other interfaces. Then, every so often, you'll implement multiple interfaces, sometimes decorating a class with new behavior, and sometimes requiring that class to implement behavior defined in the new interface. Regardless, you've got a *lot* of tools at your disposal.

But there are some rather unique cases that you'll run across every once in a while, and this one is something Kotlin is particularly interested in: the idea of an object that may be composed of more than one implementation of an interface. This gets a little tricky, so take your time, but the rewards are worth working through this use case.

Abstract Classes Move from Generic to Specific

Suppose you have a `Spy` class. This is representing a spy that has as its main function escaping from trouble (because what else do spies do?). And like all good spies, there are several ways to escape: in a car, in a motorcycle, or in a plane.

To make this possible, start out by tweaking the `Vehicle` class. Instead of a `drive()` function, it needs a simpler, more generic name: `go()`. Listing 12.26 makes this adjustment.

LISTING 12.26 A MODIFIED VEHICLE INTERFACE WITH A MORE GENERIC NAME FOR MOVING

```
package org.wiley.kotlin.vehicle

interface Vehicle {
    val maxSpeed: Int

    val purpose: String
      get() = "This interface models vehicles of all kinds"

    fun start()
    fun stop()
    fun go(direction: String, speed: Int) : Int

    fun stats(): String {
        return "This vehicle is a ${this::class.simpleName} and has a maximum speed
of ${maxSpeed}"
    }
}
```

Now this change appears to break a *lot* of code. Suddenly all those subclasses of Car need updates, because they all implement drive(), and not the new more generic go(). But this change to Vehicle and how it affects Car is actually good—the Car is an abstract class and therefore makes this a smooth fix. Check out Listing 12.27, which defines drive() as before, and now implements go() to simply call drive().

LISTING 12.27: Updating Car to pass go() calls to drive()

```
package org.wiley.kotlin.car

import org.wiley.kotlin.vehicle.Vehicle

abstract class Car(val model: String, val color: String) : Vehicle {

    override fun go(direction: String, speed: Int): Int {
        return drive(direction, speed)
    }

    abstract fun drive(direction: String, speed: Int) : Int

    override fun toString() : String {
        return "${this::class.simpleName} ${model} in ${color}"
    }
}
```

At this point, you should be able to compile all your existing classes and things will work.

> **NOTE** *One possible exception is your* Volkswagen *class, which originally (way back in Listing 12.17) implemented* Vehicle *directly. You can modify that to also extend* Car, *or change its* drive() *function to implement* go() *instead.*

You've also just seen the power of inheritance in action. A very generic interface, `Vehicle`, defined generic behavior: a `go()` function. Then, one level down, `Car` implemented `Vehicle` and moved from the generic `go()` to the more specific `drive()`. Finally, concrete classes offer behavior that backs all of that. However, at the top of the inheritance tree, it's still the `go()` function that kicked all this behavior off. You can see this illustrated in Figure 12.2.

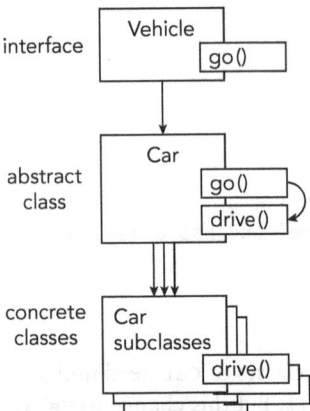

FIGURE 12.2: The go() function in Vehicle eventually is handled by drive() implementations in concrete subclasses of Car.

More Specificity Means More Inheritance

To round things out, here's a fairly significant chunk of code. Listing 12.28 defines a `Motorcycle` abstract class, and Listings 12.29 and 12.30 provide two concrete subclasses: `Harley` and `Honda` (of the motorcycle variety). For simplicity, `Motorcycle` provides implementations of `start()` and `stop()`. This is just to make the subclasses easier to code for the example's purpose.

LISTING 12.28: The Motorcycle class is a close analog to the Car class

```kotlin
package org.wiley.kotlin.motorcycle

import org.wiley.kotlin.vehicle.Vehicle

abstract class Motorcycle(model: String) : Vehicle {

    override fun start() {
        println("Starting the bike")
    }

    override fun stop() {
        println("Stopping the bike")
    }
```

```kotlin
    override fun go(direction: String, speed: Int): Int {
        return ride(direction, speed)
    }

    abstract fun ride(direction: String, speed: Int) : Int
}
```

LISTING 12.29: One concrete subclass of Motorcycle: a Harley

```kotlin
package org.wiley.kotlin.motorcycle

class Harley(model: String) : Motorcycle(model) {

    override var maxSpeed : Int = 322

    override fun ride(direction: String, speed: Int) : Int {
        println("Riding off on a ${this::class.simpleName} at ${speed} speed!")
        return speed/60
    }
}
```

LISTING 12.30: Another subclass of Motorcycle, this time a Honda

```kotlin
package org.wiley.kotlin.motorcycle

import org.wiley.kotlin.vehicle.HondaManufacturer

class Honda(model: String) : Motorcycle(model), HondaManufacturer {
    override var maxSpeed : Int = 190

    override fun ride(direction: String, speed: Int) : Int {
        println("Pushing the ${this::class.simpleName} to the limit at ${speed}
miles per hour!")
        return speed/60
    }
}
```

Now, a little more preparation work. (It will be worth it, and you can always download the code samples from the book's website to save some time.) Create a similar structure for a new vehicle type, a `Plane`. `Plane` is shown in Listing 12.31, with two variations in Listings 12.32 and 12.33.

LISTING 12.31: A Plane implementation of Vehicle

```kotlin
package org.wiley.kotlin.plane

import org.wiley.kotlin.vehicle.Vehicle

abstract class Plane(description: String) : Vehicle {
```

continues

LISTING 12.31 *(continued)*

```
    override fun start() {
        println("Starting the plane")
    }

    override fun stop() {
        println("Stopping the plane")
    }

    override fun go(direction: String, speed: Int): Int {
        return fly(direction, speed)
    }

    abstract fun fly(direction: String, speed: Int) : Int
}
```

LISTING 12.32: A Plane subclass: the B52!

```
package org.wiley.kotlin.plane

class B52() : Plane("B-52") {
    override var maxSpeed : Int = 650

    override fun fly(direction: String, speed: Int) : Int {
        println("Away we go on the B52, ${speed} miles an hour to the
${direction}!")
        return speed/60
    }
}
```

LISTING 12.33: One More Plane: a hang glider

```
package org.wiley.kotlin.plane

class HangGlider(val wingType: String) : Plane(wingType) {
    override var maxSpeed : Int = 80

    override fun fly(direction: String, speed: Int) : Int {
        println("Off in a ${wingType} glider, headed ${direction}")
        return speed/60
    }
}
```

Delegating to a Property

So, what do you do with all of these classes and interfaces? Well, take a Spy class. A good Spy really should be able to flee using all possible modes of transportation. From a certain point of view, a Spy could really implement all of these: it could have a Car implementation, a Plane implementation, and a Motorcycle implementation.

> **NOTE** *A good* Spy *would really have a boat, too, but that was just too much example code for one section!*

Listing 12.34 shows how you might create a Spy class that has several vehicles available to it.

LISTING 12.34: The Spy class, using several vehicles

```
package org.wiley.kotlin.person

import org.wiley.kotlin.car.Car
import org.wiley.kotlin.motorcycle.Motorcycle
import org.wiley.kotlin.plane.Plane

class Spy(val name: String, val car: Car, val bike: Motorcycle, val plane: Plane) {
    // Needs implementation
}
```

Now you could write a flee() method that looks something like this:

```
fun flee(vehicle: String, direction: String, speed: Int) {
    when (vehicle.toUpperCase()) {
        "PLANE" -> plane.fly(direction, speed)
        "BIKE" -> bike.ride(direction, speed)
        "CAR" -> car.drive(direction, speed)
        else -> println("I don't have that vehicle!")
    }
}
```

That's pretty sensible. It works but has some issues: you should set up an enum to deal with the different vehicle names, you are passing around a vehicle String, and it's just not ideal.

This is where delegation comes in. First, simplify Spy to take in a single Vehicle implementation. This property can store whatever vehicle the spy needs to escape:

```
class Spy(val name: String, var vehicle)
```

Now, you need a way to implement Vehicle, without implementing Vehicle. Yes, as odd as that sounded, it is exactly what's needed. You want to delegate calls to Spy that apply to the Vehicle interface through to the vehicle property. You do that using the by keyword.

Adjust Spy to look like Listing 12.35.

LISTING 12.35: Delegating Vehicle calls to a property

```
package org.wiley.kotlin.person

import org.wiley.kotlin.vehicle.Vehicle
```

continues

LISTING 12.35 *(continued)*

```
class Spy(val name: String, var vehicle: Vehicle) : Vehicle by vehicle {

    /*
    fun flee(vehicle: String, direction: String, speed: Int) {
        when (vehicle.toUpperCase()) {
            "PLANE" -> plane.fly(direction, speed)
            "BIKE" -> bike.ride(direction, speed)
            "CAR" -> car.drive(direction, speed)
            else -> println("I don't have that vehicle!")
        }
    }
    */
}
```

> **NOTE** *The original* `flee()` *function is left in and commented out so you know to remove that code.*

This tells Kotlin a couple of things:

➤   `Spy` is implementing the `Vehicle` interface.

➤   However, any calls that apply to `Vehicle`—like a reference to `maxSpeed` or the `go()` function—should be delegated to the `vehicle` property.

Now build some sample code, such as that shown in Listing 12.36, to see how this all comes together.

LISTING 12.36: A test program for the delegating Spy

```
import org.wiley.kotlin.car.BMW
import org.wiley.kotlin.motorcycle.Harley
import org.wiley.kotlin.person.Spy
import org.wiley.kotlin.plane.HangGlider

fun main() {
    val bike = Harley("Road King")
    val plane = HangGlider("Fixed Wing")
    val car = BMW("M3", "blue")

    var spy = Spy("Black Spy", bike)
    spy.go("North", 145)
}
```

Run this, and the output shows the `Spy` instance using the `bike` to travel:

```
Riding off on a Harley at 145 speed!
```

This is a remarkably elegant solution for a tricky problem. Gone is the `when` statement in the `flee()` method, worrying about `String` vehicle names, and instead, you've got a clean implementation of `Vehicle` *without* clouding up the core `Spy` functionality.

Delegation Occurs at Instantiation

Be careful, though, as there are still some tricky things to watch out for here. First, the delegation is attached to the object when it is passed into the instance of `Spy`. Look at this code:

```
var spy = Spy("Black Spy", bike)
spy.go("North", 145)
spy.vehicle = plane
spy.go("West", 280)
spy.vehicle = car
spy.go("Northwest", 128)
```

You might think that each invocation of `go()` would use the current `vehicle` property. But that's not what happens; here's the output:

```
Riding off on a Harley at 145 speed!
Riding off on a Harley at 280 speed!
Riding off on a Harley at 128 speed!
```

You'd have to actually re-create the instance of `Spy` for this to work:

```
var spy = Spy("Black Spy", bike)
spy.go("North", 145)

spy = Spy(spy.name, plane)
// spy.vehicle = plane
spy.go("West", 280)

spy = Spy(spy.name, car)
// spy.vehicle = car
spy.go("Northwest", 128)
```

This is messy, and you end up with a lot of code that is not particularly self-documenting. If this is the behavior you want, you could consider converting `Plane`, `Motorcycle`, and `Car` into interfaces themselves, each extending `Vehicle`.

> **NOTE** *There's no sample code for this conversion, as it is a great exercise for you to do. You'll get even more comfortable with Kotlin's inheritance and the relationship between abstract classes and interfaces, as well as extension versus implementation.*

If you made that change, you could update `Spy` to look like Listing 12.37, which is a clever change.

LISTING 12.37: Delegating to three different interfaces

```
package org.wiley.kotlin.person

import org.wiley.kotlin.car.Car
import org.wiley.kotlin.motorcycle.Motorcycle
import org.wiley.kotlin.plane.Plane

class Spy(val name: String, bike: Motorcycle, plane: Plane, car: Car) : Motorcycle
by bike, Plane by plane, Car by car {
}
```

Now `Spy` takes in three vehicles, and implements three interfaces, all through delegation. You can now call functions from all three interfaces on `Spy`! You could do something like this:

```
// Code that only works if delegating to three interfaces
spy = Spy("White Spy", bike, plane, car)
spy.ride("North", 145)
spy.fly("West", 280)
spy.ride("Northwest", 128)
```

This is pretty cool and shows just how flexible Kotlin will let you be with a little forethought and planning.

INHERITANCE REQUIRES FORETHOUGHT AND AFTERTHOUGHT

At this point, you've seen a lot of varieties of inheritance: interfaces, abstract classes, extension, implementation, decoration, and now delegation. In many cases, these are all options. In other words, there's rarely a time when only one type of inheritance works. In that sense, inheritance is similar to scope functions: you'll use the one that best suits your particular needs and style at the time.

However, the limitations of inheritance do bear weight. You can't extend more than one class, while you can implement multiple interfaces. You can't delegate to an abstract class, but you can to an interface implementation.

This will cause you to work and rework your inheritance tree. It's great to suppose you can always get it right by planning ahead, but sometimes you'll need to go back and change things, and those changes will give you more flexibility . . . until you need to make the next set of changes.

Regardless, you now have the tools to build rich object hierarchies and better model the objects your programs will depend upon.

13
Kotlin: The Next Step

WHAT'S IN THIS CHAPTER?

➤ Kotlin for Android

➤ Kotlin for Java

➤ Building with Gradle

➤ Start writing your own code!

PROGRAMMING KOTLIN FOR ANDROID

One of the most common uses for Kotlin is programming mobile applications, especially for the Android platform. You may be surprised that this didn't receive major coverage in this book. However, everything you do to program a mobile application depends on objects, functions, instantiation, inheritance, and the topics covered in this book. It's far more important that you have an idea of how Kotlin works than to learn a bunch of odd class names and build a mobile application—all unsure of what you're actually creating.

Kotlin for Android Is Still Just Kotlin

The following is a prerequisites notice for the simplest possible Android app that Google's Codelabs teaches you to build:

> This codelab is written for programmers and assumes that you know either the Java or Kotlin programming language. If you are an experienced programmer and adept at reading code, you will likely be able to follow this codelab, even if you don't have much experience with Kotlin.

As you can see, you're going to have to know Kotlin in general before you can get into any specific use cases, mobile or otherwise. Once you do know Kotlin, then going deeper into any area is largely centered on two areas:

➤ **Learning new libraries and possibly syntax.** This is the case for Android programming in particular: there are a *lot* of mobile-specific libraries you'll need to get used to working with.

➤ **Learning new syntax and idioms.** You'll remember the idea of programming idiomatic Kotlin from Chapter 11. This is the concept of working with libraries in a way that is accepted as "best" for a particular use case.

Listing 13.1 shows a fragment from the same Google Codelabs tutorial on starting out in Android.

LISTING 13.1: Loading and displaying text on an initial view

```kotlin
override fun onViewCreated(view: View, savedInstanceState: Bundle?) {
    super.onViewCreated(view, savedInstanceState)

    view.findViewById<Button>(R.id.random_button).setOnClickListener {
        val showCountTextView = view.findViewById<TextView>(R.id.textview_first)
        val currentCount = showCountTextView.text.toString().toInt()
        val action = FirstFragmentDirections.actionFirstFragmentToSecondFragment
(currentCount)
        findNavController().navigate(action)
    }

    // find the toast_button by its ID
    view.findViewById<Button>(R.id.toast_button).setOnClickListener {
        // create a Toast with some text, to appear for a short time
        val myToast = Toast.makeText(context, "Hello Toast!", Toast.LENGTH_SHORT)
        // show the Toast
        myToast.show()
    }

    view.findViewById<Button>(R.id.count_button).setOnClickListener {
        countMe(view)
    }
}
```

Fully two-thirds of this code is operating with Android libraries and the `android.view.View` object. This is specific to building visual mobile applications, and you'll see the `android.widget.TextView` class also appear in this listing.

These classes operate like any other Kotlin class—even though they build upon Java objects aimed at Android. Much of the code is doing the things you've learned from the earliest chapters: inheriting, calling functions on the base class, creating variables, and passing in lambdas to other functions like `setOnClickListener()`.

Move from Concept to Example

There's another important difference in how you should approach learning a skill like programming Kotlin for Android as opposed to learning Kotlin overall. As you've certainly seen by now, there are at least as many explanations of syntax in this book as there are actual code samples. Additionally, the samples in this book are relatively basic, because the focus is on concept and fundamental learning. How does inheritance work? How does an abstract class work?

The result is that the examples are simpler, and the explanation is longer. However, you are now ready to flip that and focus on longer code examples, tutorials, and hundreds and at times thousands of lines of code. You need far less explanation now, as you understand the basics. The best thing you can do is to code, code, code.

Here are a few good code-heavy tutorial sites to start with:

➤ `codelabs.developers.google.com`: Google's Codelabs has been mentioned a few times already. It's really the gold standard for getting into Android with Kotlin. It still has a fair bit of explanation but does focus on practical examples and writing a *lot* of code. It's an ideal next step after finishing this book.

➤ `developer.android.com`: Android's own developer site actually links you back to Codelabs in many cases, but organizes things a bit differently and is a different view of where to get into Android programming.

Beyond these two sites, any web search will find you a ton of good sites for getting into Android programming using Kotlin.

KOTLIN AND JAVA ARE GREAT COMPANIONS

Just as using Kotlin for Android programming is a great pairing, interacting with Java code in Kotlin is also a great combination. However, the relationship between these two pairs is a bit different.

When it comes to Android, you'll use Kotlin to interact with Android-based phones. You'll primarily write all Kotlin code, and rarely need anything beyond the Kotlin libraries that Android provides. With Java, Kotlin often exists *alongside* existing Java code, and you may often move between a piece of Java code and a piece of Kotlin code, or even take in a piece of data from Java, work on it in Kotlin, and send it back to Java. As a result, you'll likely need to have a fairly solid understanding of Java in addition to Kotlin for this use case.

Your IDE Is a Key Component

For most Kotlin applications—building mobile applications, desktop applications, or anything else—a lot of your work is going to be within a local project that is then deployed elsewhere.

> **NOTE** *If your background is in web programming, Kotlin will cause you to level up your local development skills. This is quite different than (for example) developing a React application that interacts with an API gateway and lambdas in Amazon Web Services (AWS) or similar.*
>
> *While you certainly can interact with APIs in Kotlin, and deploy to a cloud platform, many of your applications will be compiled locally and deployed as a complete unit of work. In those cases, the IDE really does become a critical part of your toolchain.*

If you get Kotlin code and Java code into the same project, they can call each other without issue. This typically involves telling your IDE that you want to use the JVM. This will vary for different IDEs, but in IntelliJ, you don't have to do anything; Java support is built in. Listing 13.2 is a simple version of User from previous chapters that uses the Java Date package without issue.

LISTING 13.2: Interacting with Java from Kotlin

```
package org.wiley.kotlin.user

import java.util.Date

data class User(var email: String, var firstName: String, var lastName: String)
{

    val createdOn = Date()

    override fun toString(): String {
        return "$firstName $lastName with an email of $email created on
${createdOn}"
    }
}
```

If you didn't know better, you'd never know that this class mixes Java and Kotlin. But java.util .Date is very much a Java package, and then in toString(), the property that stores a Date instance is treated like any other Kotlin property:

```
return "$firstName $lastName with an email of $email created on ${createdOn}"
```

Listing 13.3 shows a simple program to use this class.

LISTING 13.3: Using the Kotlin class that in turn uses Java

```
import org.wiley.kotlin.user.User

fun main() {
    var user = User("wayne.scott@example.com", "Wayne", "Scott")

    println(user)
}
```

The output is nothing remarkable at all—and that in itself is interesting:

```
Wayne Scott with an email of wayne.scott@example.com created on Thu Sep 24
08:00:32 CDT 2020
```

Kotlin Is Compiled to Bytecode for the Java Virtual Machine

There's no big secret here; rather, Kotlin code is compiled down into bytecode that can run on the Java Virtual Machine (JVM). So while it's pretty cool and unique that you can seamlessly weave Java and Kotlin into the same program, at the end of the day, you've been writing code that runs on the JVM for a solid 13 chapters now.

The benefit is that you don't need to do anything special to get interoperability with Java. You just import the needed classes from Java packages and work with them from Kotlin.

There are some additional wrinkles that you may have to get into, but they get into fairly specific use cases. The best thing to do if you want to get deeper into Java interoperability is to check out Kotlin's official documentation on the matter at `kotlinlang.org/docs/reference/java-interop.html` (for calling Java code from Kotlin) and `kotlinlang.org/docs/reference/java-to-kotlin-interop.html` (for calling Kotlin code from Java). Both documents will give you more specifics.

Gradle Gives You Project Build Capabilities

If you're looking to build a project outside of your IDE, you can use Gradle (`gradle.org`) to build a Kotlin project that also references Java code. Gradle allows you to specify the versions of Kotlin and optionally Java that you want to use and lock your project to those versions.

You can also use Gradle to specify external dependencies, indicate the platforms you want to build for (which may use different versions of the JVM), and set properties used as part of compilation and structuring of your project. Once again, the Kotlin documentation is a good place to start: `kotlinlang.org/docs/reference/using-gradle.html`.

WHEN KOTLIN QUESTIONS STILL EXIST

If you're looking for more details on a particular subject, your first stop should be the Kotlin reference documentation hosted at `kotlinlang.org/docs/reference`. Figure 13.1 shows you the landing page.

You can find almost every topic you could imagine along the left-hand menu. You'll also find a lot of reference code in this documentation set. However, at times the examples are a bit academic and less clear. Here's an example:

```
inline fun <reified A, reified B> Pair<*, *>.asPairOf(): Pair<A, B>? {
    if (first !is A || second !is B) return null
    return first as A to second as B
}
```

While this is accurate, you're going to have to spend some time reading the text slowly to capture all the detail. If you need a deeper explanation on a topic covered in this book, the reference documentation is a great place to begin.

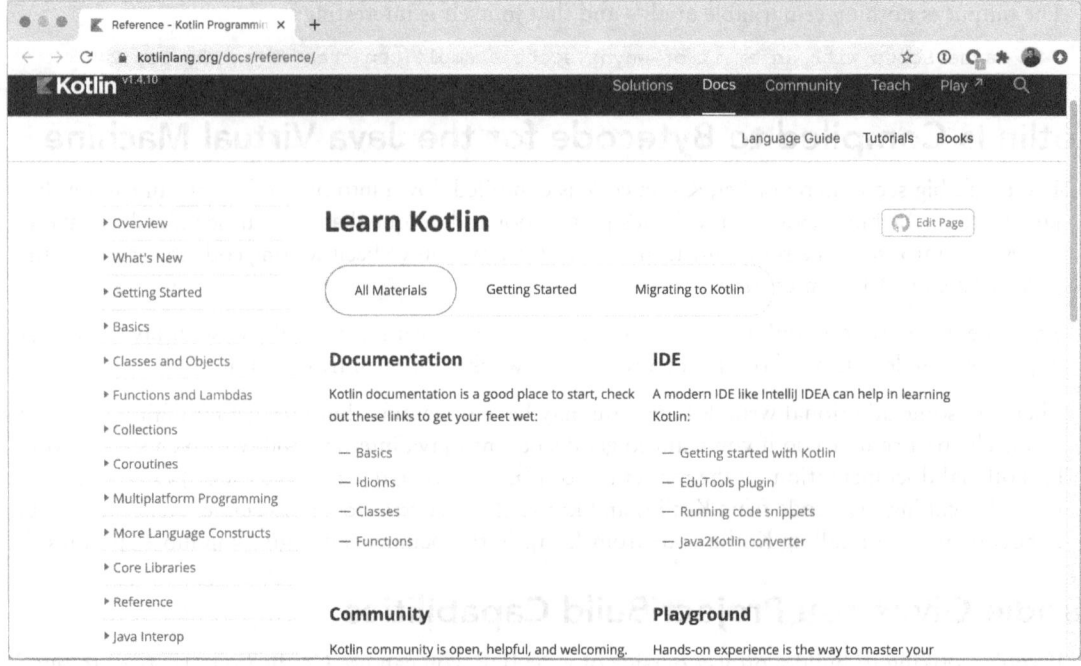

FIGURE 13.1: Kotlin's reference documentation is thorough and helpful.

Use the Internet to Supplement Your Own Needs and Learning Style

If you want more than the reference documentation provides, there are some tried and true online reference sites that are consistently helpful:

➤ baeldung.com: Like the reference documentation, baeldung.com is a bit of an academic-leaning site. You'll get a lot of programming theory, and lots of generic types. However, you'll also get some really interesting articles on design patterns and using Kotlin to its fullest.

➤ medium.com: The site medium.com is a great go-to for a *lot* of topics, from culture to programming. Kotlin is covered a bit sporadically, and you might find three articles on inheritance and nothing on mobile at all, but it's still a great site when you find what you're looking for. It's also very practical, with articles written for programmers that may *not* have as much interest in the science part of computer science.

➤ journaldev.com: The site journaldev.com is another site that has varied coverage, but is a great site for practical articles that have some pretty clear examples.

➤ play.kotlinlang.org/koans/overview: The Kotlin Koans are a pretty unique little playground. It's an interactive way to learn as you go. It's structured but almost entirely experiential, so if you want a lot of clear and explicit explanation, this might not be the site for you. Still, it's worth checking out. Figure 13.2 shows an example of a Koan on data classes that requires you to fill out missing code with some hints along the right.

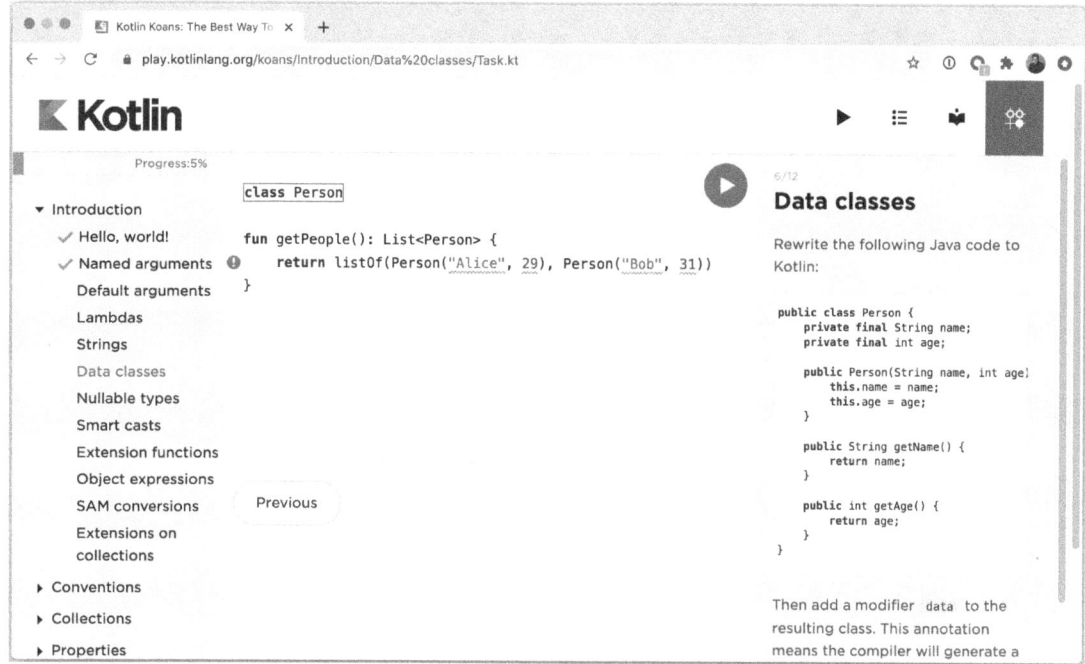

FIGURE 13.2: The Kotlin Koans aren't for everyone, but they're great for interactive learning.

NOW WHAT?

With an entire book on Kotlin now in your rearview mirror, you are more than ready to take what you've learned and build your own programs, classes, objects, functions, lambdas, flows, and applications. You're also ready to make that shift discussed earlier in this chapter from reading about code to coding. Yes, keep this book and your browser with Kotlin documentation handy, but get into your IDE. You'll find you may never look back.

Enjoy, and happy programming!

INDEX